SpringerBriefs in Computer Science

SpringerBriefs present concise summaries of cutting-edge research and practical applications across a wide spectrum of fields. Featuring compact volumes of 50 to 125 pages, the series covers a range of content from professional to academic.

Typical topics might include:

- A timely report of state-of-the art analytical techniques
- A bridge between new research results, as published in journal articles, and a contextual literature review
- A snapshot of a hot or emerging topic
- An in-depth case study or clinical example
- A presentation of core concepts that students must understand in order to make independent contributions

Briefs allow authors to present their ideas and readers to absorb them with minimal time investment. Briefs will be published as part of Springer's eBook collection, with millions of users worldwide. In addition, Briefs will be available for individual print and electronic purchase. Briefs are characterized by fast, global electronic dissemination, standard publishing contracts, easy-to-use manuscript preparation and formatting guidelines, and expedited production schedules. We aim for publication 8–12 weeks after acceptance. Both solicited and unsolicited manuscripts are considered for publication in this series.

Indexing: This series is indexed in Scopus, Ei-Compendex, and zbMATH

More information about this series at http://www.springer.com/series/10028

Ahmad A. Aziz El-Banna • Kaishun Wu

Machine Learning Modeling for IoUT Networks

Internet of Underwater Things

Springer

Ahmad A. Aziz El-Banna
Shenzhen University
Nanshan, China

Benha University (On leave)
Benha, Egypt

Kaishun Wu
Shenzhen University
Nanshan, China

ISSN 2191-5768   ISSN 2191-5776   (electronic)
SpringerBriefs in Computer Science
ISBN 978-3-030-68566-9   ISBN 978-3-030-68567-6   (eBook)
https://doi.org/10.1007/978-3-030-68567-6

This Springer imprint is published by the registered company Springer Nature Switzerland AG
The registered company address is: Gewerbestrasse 11, 6330 Cham, Switzerland

To our fantastic families, thank you for the continuous and favorable support

Preface

Our objective of writing this book is to draw some outlines to enable the reader to find a good source of knowledge that could help them to understand applied technologies in underwater communication as well as the existing challenges and some suggested solutions.

The first aim of this book is to shed light on the leading variable physical properties of water that affect the transmission of the core carrier for underwater communications, i.e., acoustic waves, and what leads to changes in its characteristics, e.g., intensity, resulting in inaccurate production of the deployed technologies that build on the assumption of fixed speed transmission of sound in underwater environments.

The second major aim is to investigate the application of machine learning (ML) techniques in settling diverse challenges that are encountered during deployment of underwater technologies, such as the Internet of Underwater Things (IoUT) and multi-modal underwater networks, and to capitalize on their merits. In addition, ML has the capabilities to treat the traditional underwater model-driven problems by considering the enormous measured data into appropriate data-driven problems and handle them to design a proper and adaptive behavioral modeling of these problems. This overcomes the main underwater problem where there is still no generic model that exists for the underwater environments because of the extremely harsh and fluctuating nature of such ambiance over the spatio-temporal domains since the ML techniques, unlike the theoretical systems, do not rely on explicit or certain propagation models or assumptions.

The book provides IoUT network and node structure and the ML modeling for underwater communication in Chap. 1, the key physical variables of water and their interrelationships in Chap. 2, an example of channel modeling and an adaptive transmission framework for underwater networks in Chap. 3, two examples of the positioning systems in Chap. 4, and application of the decision tree as a classifier and dynamic modeling using neural networks for underwater techniques. The book also proposes in Chap. 6 some challenges faced by underwater communication and some glimpses of the solutions from both communication and data science perspectives.

Finally, we hope this book satisfies the need of the community members and they find it interesting and fruitful.

Nanshan, China   Ahmad A. Aziz El-Banna

Nanshan, China   Kaishun Wu

Contents

Acronyms

ADCs	Analog-to-Digital Converters
AF	Amplify-and-Forward
ANNs	Artificial Neural Networks
AoA	Angle of Arrival
AWGN	Additive White Gaussian Noise
BER	Bit-Error-Rate
Bval	Best Validation
CFNN	Cascade-Forward Neural Network
CM	Confidence Metric
DDNN	Distributed Delay Neural Network
DF	Decode-and-Forward
DNN	Deep Neural Network
DT	Decision Tree
DyNets	Dynamic Neural Networks
FFNN	Feed-Forward Neural Network
GPS	Global System Positioning
IoT	Internet of Things
IoUT	Internet ofUnderwater Things
KPV	Key Physical Variables
ML	Machine Learning
MSE	Mean Squared Normalized Error
NAR	Nonlinear Autoregressive
NARX	Nonlinear Autoregressive Exogenous
NNs	Neural Networks
pH	Power of Hydrogen
PSK	Phase Shift Keying
QAM	Quadrature Amplitude Modulation
RNN	Recurrent Neural Network
ROV	Remotely Operated Underwater Vehicle
RSS	Received Signal Strength
RSSI	Received Signal Strength Indicator

SER   Symbol Error Rate
TDLs   Tapped Delay Lines
TDNN   Time Delay Neural Network
TDoA   Time Difference of Arrival
ToA   Time of Arrival
TPLs   Transmission Power Levels
UWSNs   Underwater Wireless Sensor Networks

Chapter 1
Introduction to Underwater Communication and IoUT Networks

1.1 Underwater Communication

Since no typical underwater environment exists, there is also no existence of a typical communication channel, and that is considered the fundamental challenge for underwater transmissions. Dramatic water dynamics and air bubbles along with strong signals absorption and considerable sources of scattering are some reasons that complicate underwater communication. Other reasons are the diverse underwater environments that extend from rivers, lakes, seas, to oceans, and every environment has its specific physical properties which usually change from a region to another and from season to season and even from day to night. Some examples for these physical variables are water temperature, salinity, pressure and density, power of hydrogen (pH), water conductivity, and wave speed. The aforementioned reasons lead to many challenges that should be considered in the design of underwater wireless sensor networks (UWSNs) [5, 7]. One challenge is the network requirements in terms of throughput and energy efficiency especially for the battery-based nodes that are used to collect data and monitor the underwater environment. Another challenge is that the protocols designed for terrestrial wireless communication usually are not applicable for such a harsh environment [23]; therefore, these protocols should be adapted to fulfill the UWSN requirements.

Furthermore, high transmission error rates are considered one of the critical restrictions that face UWSNs [5] due to the dynamic topology of UWSNs, e.g., the passive movements of the sensor nodes by the water currents and the persistent movements of the sea organisms [22]. Therefore, we need an adaptive transmission system to utilize the good communication opportunities of the environment that surrounds the sensor nodes to achieve a distinct performance of the underwater networks. Moreover, another form of transmission, called cooperative transmission, can mitigate the defiance of high error rates via providing some form of diversity by employing an intermediate relay node between the source and the destination nodes to improve the spectral efficiency [3]. For example, amplify-and-forward (AF) and

© The Author(s), under exclusive license to Springer Nature Switzerland AG 2021
A. A. Aziz El-Banna, K. Wu, *Machine Learning Modeling for IoUT Networks*,
SpringerBriefs in Computer Science, https://doi.org/10.1007/978-3-030-68567-6_1

decode-and-forward (DF) are two traditional relaying schemes [12], but each one of them has its own advantages and disadvantages, i.e., analysis simplicity and low complexity but with poor performance under many scenarios of the former and higher performance but with a considerable computational complexity of the later. However, hybrid relaying protocols, e.g., [3, 12, 13], could make use of the merits of AF and DF based on transmission circumstances and are considered a proper compromise of the two conventional schemes. Nevertheless, whether to use either AF or DF is still a challenging task and needs a proper selection criterion to indicate the method of choice between them to improve the transmission under different channel states.

Many studies have recently proposed various hybrid and cooperative transmission schemes for different underwater environments, e.g., [22, 23] and [4, 11, 15, 17, 18, 20, 21], by considering the two trends of optimal relay selection and routing protocols. However, most of the work done before [22] assumed known locations of all the network nodes which is a defy task for UWSNs. Moreover, and although employing only local depth information without the need of the full-dimensional location information of the nodes besides considering the residual relay energy in some work, various studies didn't consider other factors that affect the underwater transmitted signal and could lead to bad states of the communication channel.

On the other hand, underwater sensor nodes often employ numerous types of sensors and actuators to monitor and control the behavior of the motes and other device components besides the outer-environmental characteristics. However, almost all of the underwater sensor nodes are equipped with a tinny and inexpensive temperature sensor plus many other data-loggers that are used to monitor the water characteristics. Therefore, one promising solution is developing schemes that utilize the collected data from such sensors and data-loggers to compute a selection metric to adaptively choose the suitable networking settings such as relaying schemes, for example, AF or DF is currently suitable to exploit it in transmission to mitigate any bad conditions that happen in the communication medium. Moreover, various wireless sensor motes employ numerous transmission power levels (TPLs) and modulation methods [24] that could be activated to control the transmission power besides the data rate. An environmental-based selection metric is proposed in [6] to indicate the suitable relaying scheme and power levels TPLs besides adapting the modulation methods used in transmission.

1.2 IoUT Network and Node Structure

Various underwater environments are promising areas to deploy recent innovative applications and technologies such as IoUT and multimodal underwater networks. IoUT is a recent category of the Internet of Things (IoT) technology that extends the operation of UWSNs [9]. In addition, IoUT could employ cloud computing platforms to assist in the communication process between other components of the network. Moreover, IoUT is expected to be extensively deployed in various underwater environments to cover numerous underwater monitoring and actuation applications toward smart worldwide networking of underwater devices [2, 5].

1.2.1 Network Architecture

Figure 1.1 depicts a typical architecture of an IoUT network, where multiple sensor nodes are scattered randomly and exist at depths z_R and are separated from each other by Euclidean distances d in a certain area to perform their tasks. The nodes may be static, semi-autonomous, or autonomous nodes (or vehicles) and are connected to the IoUT cloud using a surface gateway. The surface sink node (may also be a buoyage or a ship) is often equipped with both acoustic and radio modems, where the acoustic link is used for underwater communication between underwater nodes, divers, and submarines, while the RF link is used for terrestrial and satellite communication.

For multihop communication between the source node (S) and the destination node (D), the relay nodes (R) are scattered over n-tires between S and D, where each

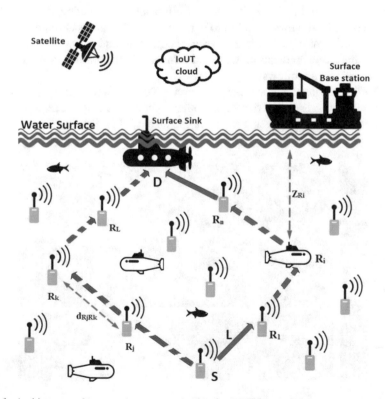

Fig. 1.1 Architecture of cooperative communication for IoUT Network

tire contains k-relays, and each tire is considered one hop in the message-forwarding path assuming the transmission is performed over n-communication phases $\phi_1 : \phi_n$ in every route.

In Fig. 1.1, we consider a cooperative transmission scenario between the deployed sensor nodes where one sensor node broadcasts its data packets to its neighboring nodes but only some nodes decide to process the received packets according to predetermined criteria, e.g., the transmission link quality or the instantaneously available resources of the relay node. Moreover, any cooperating relay node can also choose to just amplify then forward or decode then forward the received signal according to that specified metric.

The signal received at the relay or that combined at the destination is given by the general form $Y = HX + N$, where X and Y are the transmitted and received signals, respectively, and H and N are the channel responses and noise affecting the transmission, respectively.

1.2.2   Sensor Node Architecture

Figure 1.2 shows the main components of a typical sensor node (or mote). Many layers are combined within the chip to perform the different tasks required by each node to attain its assigned task. However, the number and types of layers clearly depend on the target application; we could summarize the common components and layers of the mote as follows.

At the bottom layer lies the power source module where rechargeable or non-rechargeable batteries are used to supply other components with the required energy. Some sorts of energy harvesting modules can also exist at that layer, where energy

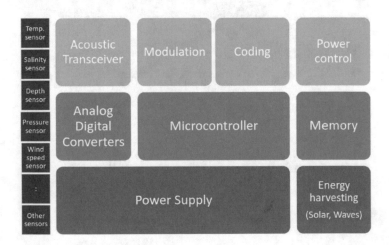

Fig. 1.2 Main components of an UWSN node

can be harvested in the underwater environment from the available ambient sources such as solar energy and water currents.

A second layer includes the controller which is the brain and dispatcher of all the tasks performed by the node, the memory to store the required routines and algorithms, and the analog-to-digital converters (ADCs) which are used to sample and digitize the collected data come from the sensors layer. This sensors layer includes various types of sensors to measure and monitor the required events and the underwater characteristics such as temperature, salinity, water depth, pressure, internal waves, etc. and also contains various types of actuators such as motors.

Another layer involves the communication modules such as the modulator, channel encoder/decoder, power control unit, and the acoustic transceiver to transmit and receive the processed messages. The communication chain blocks lie in this layer, and the networking subroutines and/or routing tables are performed by the modules implemented in this layer as well.

Moreover, and depending on the target application, the node can be equipped with further modules such as microphones, cameras, and their adjacent audio/image processing units. Furthermore, global system positioning (GPS) modules may be included in some nodes; however, the cost of such nodes becomes higher than the typical ones.

1.3 Machine Learning Modeling for Underwater Communication

Machine learning (ML) techniques acquired much interest these days in the wireless communication field due to their capabilities to deduce proper solutions for numerous communication problems [1]. In addition, ML techniques could provide intelligent functions that adaptively exploit the wireless resources, optimize the network operation, and guarantee the QoS needs in real-time applications [14, 19].

Some of the ML merits

1. can enhance the performance by finding solutions faster than heuristic models
2. could be implemented with lower complexities compared to complex mathematical models
3. are able to cover other scenarios similar to those given to it during the training phase.

Moreover, there are many approaches for machine learning algorithms, but generally they are classified into supervised, unsupervised, and reinforcement learning. Figure 1.3 shows the types of the main ML techniques and some examples of their sub-types.

Supervised learning algorithms aim to produce a mathematical model, containing the inputs and the desired outputs, of a certain training data [16]. Otherwise, unsupervised learning learns to recognize the commonalities in the data to find a proper

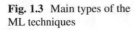

Fig. 1.3 Main types of the ML techniques

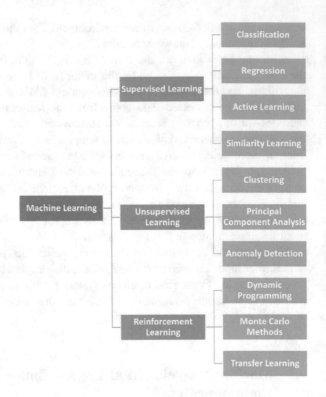

structure and hence, doesn't require examples of the output (i.e., labels). Finally, reinforcement learning doesn't require the knowledge of the labeled input/output pairs and aims to find a proper balance between the current knowledge and the uncharted one by maximizing the notion of cumulative reward [8].

Some types of algorithms used for:

1. *supervised learning*: are support vector machines, linear regression, logistic regression, naive Bayes, linear discriminant analysis, decision trees, k-nearest neighbor algorithm, neural networks, and similarity learning.
2. *unsupervised learning*: are k-means, mixture models, OPTICS algorithm, neural networks, autoencoders, deep belief nets, Hebbian learning, and generative adversarial networks.
3. *reinforcement learning*: are Monte Carlo, Q-learning, SARSA, Lambda, and DQN.

Various types of the aforementioned ML techniques and algorithms can be employed in different communication networks to improve the routing, resource allocations, channel assignments, rate adaptation, energy efficiency, etc. [10].

Classification is one of the ML tasks where its objective is to predict the target class for each case of the dataset. Several classification approaches have been introduced to solve communication problems. Recently, classification techniques such as decision tree, rule-based method, memory-based learning, Bayesian networks, neural networks, and support vector machines have been used for numerous communication problems.

Regression is another class that is extensively used in communication problems. Conventional regression techniques are basically used to solve problems where the goal is to predict infinite values using regression analysis to estimate the relationship between two or more variables. However, recent deep neural network (DNN) models outperformed state of the art in many tasks such as classification, object detection, and regression. The main framework of such models is based on stacking various layers (convolutional, pooling, or recurrent layers) followed by fully connected layers tailored based on the required task.

References

1. M.A. Alsheikh, S. Lin, D. Niyato, H. Tan, Machine learning in wireless sensor networks: Algorithms, strategies, and applications. IEEE Commun. Surv. Tutorials **16**(4), 1996–2018 (2014)
2. A.A. Aziz El-Banna, A.B. Zaky, B.M. ElHalawany, J. Zhexue Huang, K. Wu, Machine learning based dynamic cooperative transmission framework for IoUT networks, in *2019 16th Annual IEEE International Conference on Sensing, Communication, and Networking (SECON)* (2019), pp. 1–9
3. B. Can, H. Yomo, E.D. Carvalho, Hybrid forwarding scheme for cooperative relaying in OFDM based networks, in *2006 IEEE International Conference on Communications*, vol. 10, pp. 4520–4525 (2006). https://doi.org/10.1109/ICC.2006.255351
4. Y. Chen, Z. Wang, L. Wan, H. Zhou, S. Zhou, X. Xu, OFDM-modulated dynamic coded cooperation in underwater acoustic channels. IEEE J. Ocean. Eng. **40**(1), 159–168 (2015). https://doi.org/10.1109/JOE.2014.2304254
5. R.W.L. Coutinho, A. Boukerche, L.F.M. Vieira, A.A.F. Loureiro, Underwater wireless sensor networks: A new challenge for topology control-based systems. ACM Comput. Surv. **51**(1), 19:1–19:36 (2018). https://doi.org/10.1145/3154834
6. A.A.A. El-Banna, K. Wu, B.M. ElHalawany, Opportunistic cooperative transmission for underwater communication based on the water's key physical variables. IEEE Sensors J. **20**(5), 2792–2802 (2020)
7. J. Heidemann, Y. Wei, J. Wills, A. Syed, Y. Li, Research challenges and applications for underwater sensor networking, in *IEEE Wireless Communications and Networking Conference, 2006. WCNC 2006*, vol. 1 (2006), pp. 228–235
8. L.P. Kaelbling, M.L. Littman, A.W. Moore, Reinforcement learning: A survey. J. Artif. Intell. Res. **4**, 237–285 (1996)
9. C.C. Kao, Y.S. Lin, G.D. Wu, C.J. Huang, A comprehensive study on the internet of underwater things: Applications, challenges, and channel models. Sensors **17**(7), 1477 (2017). https://doi.org/10.3390/s17071477
10. S. Karunaratne, H. Gacanin, An overview of machine learning approaches in wireless mesh networks. IEEE Commun. Mag. **57**(4), 102–108 (2019)
11. A. Khan, I. Ali, A.U. Rahman, M. Imran, H. Mahmood, Co-EEORS: Cooperative energy efficient optimal relay selection protocol for underwater wireless sensor networks. IEEE Access **6**, 28777–28789 (2018). https://doi.org/10.1109/ACCESS.2018.2837108
12. J.N. Laneman, D.N.C. Tse, G.W. Wornell, Cooperative diversity in wireless networks: Efficient protocols and outage behavior. IEEE Trans. Inf. Theory **50**(12), 3062–3080 (2004). https://doi.org/10.1109/TIT.2004.838089

13. T. Liu, L. Song, Y. Li, Q. Huo, B. Jiao, Performance analysis of hybrid relay selection in cooperative wireless systems. IEEE Trans. Commun. **60**(3), 779–788 (2012). https://doi.org/10.1109/TCOMM.2012.011312.110015

14. Q. Mao, F. Hu, Q. Hao, Deep learning for intelligent wireless networks: A comprehensive survey. IEEE Commun. Surv. Tutorials **20**(4), 2595–2621 (2018)

15. H. Nasir, N. Javaid, H. Ashraf, S. Manzoor, Z.A. Khan, U. Qasim, M. Sher, CoDBR: Cooperative depth based routing for underwater wireless sensor networks, in *2014 Ninth International Conference on Broadband and Wireless Computing, Communication and Applications* (2014), pp. 52–57. https://doi.org/10.1109/BWCCA.2014.45

16. S. Russell, P. Norvig, *Artificial Intelligence: a Modern Approach*, 4th edn. (Pearson, 2020). https://www.cin.ufpe.br/~tfl2/artificial-intelligence-modern-approach.9780131038059.25368.pdf

17. P. Wang, X. Zhang, Energy-efficient relay selection for QoS provisioning in MIMO-based underwater acoustic cooperative wireless sensor networks, in *2013 47th Annual Conference on Information Sciences and Systems (CISS)* (2013), pp. 1–6. https://doi.org/10.1109/CISS.2013.6624269

18. P. Wang, L. Zhang, V.O.K. Li, Asynchronous cooperative transmission for three-dimensional underwater acoustic networks. IET Commun. **7**(4), 286–294 (2013). https://doi.org/10.1049/iet-com.2012.0314

19. T. Wang, C.K. Wen, H. Wang, F. Gao, T. Jiang, S. Jin, Deep learning for wireless physical layer: Opportunities and challenges. China Commun. **14**(11), 92–111 (2017)

20. P. Xie, J.H. Cui, L. Lao, VBF: Vector-based forwarding protocol for underwater sensor networks, in *NETWORKING 2006. Networking Technologies, Services, and Protocols; Performance of Computer and Communication Networks; Mobile and Wireless Communications Systems*, ed. by F. Boavida, T. Plagemann, B. Stiller, C. Westphal, E. Monteiro (Springer Berlin Heidelberg, Berlin, Heidelberg, 2006), pp. 1216–1221

21. F. Xu, L. Yang, Two-way relay underwater acoustic communications with multiuser decision-feedback detection and relay preprocessing, in *2012 1st IEEE International Conference on Communications in China (ICCC)* (2012), pp. 602–607. https://doi.org/10.1109/ICCChina.2012.6356955

22. H. Yan, Z.J. Shi, J.H. Cui, D B R: Depth-based routing for underwater sensor networks, in *NETWORKING 2008 Ad Hoc and Sensor Networks, Wireless Networks, Next Generation Internet*, ed. by A. Das, H.K. Pung, F.B.S. Lee, L.W.C. Wong (Springer Berlin Heidelberg, Berlin, Heidelberg, 2008), pp. 72–86

23. H. Yang, B. Liu, F. Ren, H. Wen, C. Lin, Optimization of energy efficient transmission in underwater sensor networks, in *GLOBECOM 2009 - 2009 IEEE Global Telecommunications Conference* (2009), pp. 1–6. https://doi.org/10.1109/GLOCOM.2009.5425484

24. H.U. Yildiz, B. Tavli, H. Yanikomeroglu, Transmission power control for link-level handshaking in wireless sensor networks. IEEE Sensors J. **16**(2), 561–576 (2016). https://doi.org/10.1109/JSEN.2015.2486960

Chapter 2
Seawater's Key Physical Variables

2.1 Key Physical Variables (KPVs)

2.1.1 Temperature

Temperature is the first key parameter, where it is considered a significant factor because of its twofold effect: firstly, temperature value varies around the day, i.e., it has several values at different times in the evening, night, morning, and noon, and that mutates the underwater physical properties which could affect the sound signal intensity, and secondly, all chemical processes are temperature-dependent equilibrium reactions, e.g., salinity [5], and we claim that water temperature almost affects all other water KPVs [2]. Nevertheless, the temperature of the seawater is affected by some factors such as the sunlight, the atmospheric heat transfer, turbidity, and water depth. Furthermore, underwater temperatures change from place to place and from season to another.

Figure 2.1 shows an example of a temperature-depth ocean water profile at a surface temperature of 20 °C. Note that a relation between a certain physical parameter and the water depth is called a profile. In addition, Fig. 2.2 shows the growth and decay of the seasonal temperature profiles in the upper 100 m of the ocean at one station in different months of the year (Northern Hemisphere) presented in [10].

Moreover, a typical temperature-depth ocean water profile is shown in Fig. 2.3a as reported by Fig. 1 in [9]. This temperature profile demonstrates that shallow water temperature has a large gradient and this affects numerous applications deployed in the sea surface, e.g., sea sports and water quality monitoring, but also deep water is still affected in a small manner based on the water depth.

A. A. Aziz El-Banna, K. Wu, *Machine Learning Modeling for IoUT Networks*,
SpringerBriefs in Computer Science, https://doi.org/10.1007/978-3-030-68567-6_2

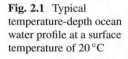

Fig. 2.1 Typical
temperature-depth ocean
water profile at a surface
temperature of 20 °C

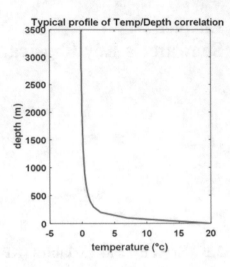

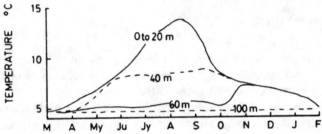

Fig. 2.2 Growth and decay of the seasonal temperature profiles in the upper 100 m of the ocean at one station in different months of the year (Northern Hemisphere) presented in [10]

2.1.2   Salinity

A second parameter is a salinity which is a measure of the "saltiness" of the water, and its importance comes from that it, along with temperature, defines the density level of seawater [9, 10] and thus the level of underwater sound speed. A typical water salinity profile is shown in Fig. 2.3b which infers that at different depths, the general features of the salinity profile remain the same but the quantitative aspects change. One method to estimate the salinity level employs the measurements of the electrical conductivity of the water; however, the seawater conductivity is highly temperature-dependent and mildly pressure-dependent [9], and from this point, we can consider the pressure as a candidate KPV.

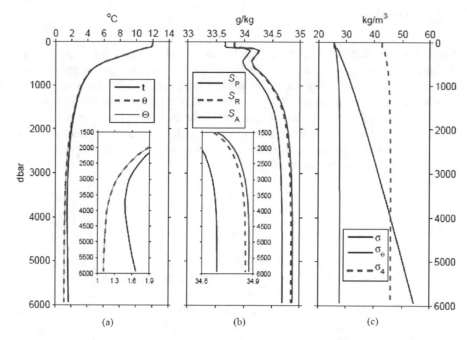

Fig. 2.3 A vertical profile of temperature and salinity at 39° N, 152° W in the North Pacific (data courtesy of the CLIVAR and Carbon Hydrographic Data Office, https://cchdo.ucsd.edu). (**a**) Vertical profiles of in situ temperature t, potential temperature θ, and conservative temperature Θ. Inset shows an expanded view of deep ocean values. (**b**) Vertical profiles of practical salinity S_P (with no units), reference salinity S_R, and absolute salinity S_A. (**c**) Vertical profiles of in situ density (σ), potential density referenced to the sea surface (σ_θ), and potential density referenced to 4000 dbar (σ_4). Reported by Fig. 1 in [9]

2.1.3 Pressure

Seawater pressure can be defined as the force exerted by a water column per unit area, and this force increases when the height of the water column increases, i.e., when we go deeper under the sea, so pressure by definition is a direct dependency on the water depth, so the pressure and depth are considered two faces of one coin, and hence we employed the depth as the third KPV in our proposed scheme. Besides the water depth, the pressure in a liquid depends on the density of that liquid, and that gives a sign for the next KPV, i.e., density.

Furthermore, we would like to clarify that the speed of sound in water increases proportionally with the increase in water temperature, salinity, and pressure (depth). The approximate change in the speed of sound with a change in each property is as follows [12]: 1 °C increase in temperature increases the sound speed by 4.0 m/s, and a growth of 1 PSU (practical salinity unit) in salinity increases the sound by 1.4 m/s, while the sound speed boosts by 17 m/s for each 1 km increase in depth (pressure).

2.1.4   Density

Density is the fourth KPV of interest; it is considered the most dominant thermo-dynamic property of seawater for oceanic circulation studies [9]. Density depends on heat content and salinity and varies slightly with pressure as mentioned above. Moreover, more saline waters at a particular temperature are denser. The density profile is shown in Fig. 2.3c which also shows the pressure effects on the density profile. The density has an inverse relationship with the speed of sound and as a result of the transmitted signal intensity, i.e., dense water causes the sound to travel at a slower rate and with less intensity.

2.1.5   pH

Additionally, power of hydrogen (pH) could be chosen as a key parameter as it follows density independently of what controls it [7], and then a perturbation in pH is considered an estimate of the density variations. It worth noting that density can be changed in two ways: one way is by heating and this significantly affects the sound, and the other way is by pressure and this doesn't affect the sound in a substantial way. However, when the density changes, it directly impacts the intensity of the underwater transmitted sound signal. The pH is used to specify the acidity or basicity of an aqueous solution and is defined as the decimal logarithm of the reciprocal of the hydrogen ion activity, (aH+), in a solution and is given as [4]

$$pH = -10\log(aH+)$$

Furthermore, temperature plays a substantial role in pH measurements. The two quantities are inversely proportional as shown in Fig. 2.4 which shows the typical values for the water pH at different temperatures [1]. In addition, the underwater pH values also change from place to place and from time to time as well as from depth to depth as shown in Fig. 2.5, and as reported by Fig. 3 in [11], and Fig. 2 and Table 1 in [8].

2.1.6   Internal Waves

We could also consider the internal waves as one of the parameters that have a great effect on underwater communication as they complicate the propagation process of sound in the water. These waves are considered the main cause of disturbances in sound speed and are the significant parameter in sea frequency spectrum changes [13]. The variation of sound speed due to internal wave movements is almost equal

Fig. 2.4 Typical pH values of water at various temperature degrees

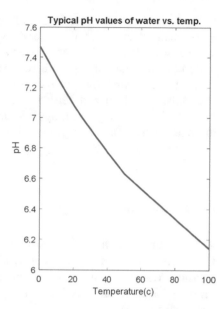

Fig. 2.5 Seawater-surface pH data collected during the first week of RRS Discovery cruise D366 as part of the UK Ocean Acidification Research Programme and presented in [11]

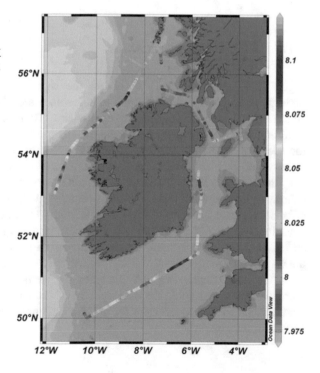

to the vertical movement of water multiplied by the differentiation of the speed of non-turbulent sound w.r.t. the water depth. A study proposed in [13] shows that the internal wave floating frequency affects the sound wave frequency and concludes that in the first 20 m of depth where the sound speed decrease rapidly, the floating frequency is maximum because of some formation principals of the water masses and density gradient and consequently the highest disturbances in sound speed happen. The density gradient by depth increasing is weakening, and consequently, the floating frequency and generated disturbances in sound speed are minimized through the internal waves.

2.1.7   Conductivity

A final parameter that has a strong correlation with salinity is conductivity. Conductivity is usually used in estimating salinity values because it is easier to measure. Moreover, it is considered an early indicator of water system changes. On the other hand, conductivity fluctuates by the change in temperature, water flow, and water levels [3, 6].

2.1.8   KPV Interrelationships

Figure 2.6 depicts a chart that bound the significant factors that change the underwater sound speed profile and as a result, affect the underwater transmitted signal intensity and their coherence relationships. By inspecting these relationships, it is found that the temperature affects the pH of water, conductivity, salinity, and the speed of sound. On the other hand, the water depth impacts the pH of water, salinity, and pressure. And in turn, water conductivity and salinity besides pressure can directly change the sound speed profile. Moreover, a change in pH and/or salinity influences the metal solubility, producing a change in density which in turn affects the signal intensity directly and indirectly by affecting the speed of sound. Additionally, there is a correlation between salinity and pH of water. Finally, wind speed, or water wave dynamics disturbs the speed of sound in the upper layers of the water.

In the next chapter, we will illustrate the energy model for the underwater communication that guide the employment of the KPV in determining a suitable confidence metric as an indicator for the channel quality and hence determining the suitable transmission settings.

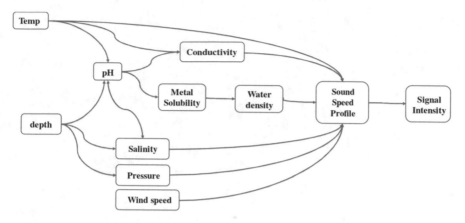

Fig. 2.6 Interrelationships between underwater parameters

References

1. A.A. Aziz El-Banna, A.B. Zaky, B.M. ElHalawany, J. Zhexue Huang, K. Wu, Machine learning based dynamic cooperative transmission framework for IoUT networks, in *2019 16th Annual IEEE International Conference on Sensing, Communication, and Networking (SECON)* (2019), pp. 1–9
2. Biscayne bay water watch (BBWW) project, field-based data documents: pH, salinity and temperature. https://sfyl.ifas.ufl.edu/media/sfylifasufledu/miami-dade/documents/sea-grant/Temperature,-Salinity-and-pH.pdf. Accessed 15 Jul 2019
3. Conductivity, salinity and total dissolved solids. https://www.fondriest.com/environmental-measurements/parameters/water-quality/conductivity-salinity-tds/
4. A.K. Covington, R. Bates, R. Durst, Definition of pH scales, standard reference values, measurement of pH and related terminology. Pure Appl. Chem. **57**(3), 531–542 (1985)
5. Critical factors affecting pH measurement. http://www.jumo.pl/pl_PL/prasa/white-papers/index.html. Accessed 15 Jul 2019
6. M. Hayashi, Temperature-electrical conductivity relation of water for environmental monitoring and geophysical data inversion. Environ. Monit. Assessm. **96**(1), 119 (2004)
7. J.M. Hernández-Ayón, A. Zirino, S. Marinone, R. Canino-Herrera, M.S. Galindo-Bect, pH-density relationships in seawater. Ciencias Marinas **29**(4), 497–508 (2003)
8. G.E. Hofmann, J.E. Smith, K.S. Johnson, U. Send, L.A. Levin, F. Micheli, A. Paytan, N.N. Price, B. Peterson, Y. Takeshita et al., High-frequency dynamics of ocean ph: a multi-ecosystem comparison. PloS One **6**(12), e28983 (2011)
9. R. Pawlowicz, Key physical variables in the ocean: Temperature, salinity, and density. Nat. Educ. Knowl. **4**(4) (2013). https://www.nature.com/scitable/knowledge/library/key-physical-variables-in-the-ocean-temperature-102805293
10. A. Paytan, Chemistry course. Class lecture, topic: "temperature, salinity, density and ocean circulation" (2006). http://ocean.stanford.edu/courses/bomc/chem/lecture_03.pdf
11. V.M. Rérolle, C.F. Floquet, M.C. Mowlem, D.P. Connelly, E.P. Achterberg, R.R. Bellerby, Seawater-ph measurements for ocean-acidification observations. TrAC Trends Anal. Chem. **40**, 146–157 (2012)
12. Science of sound tutorial. https://dosits.org/tutorials/science/tutorial-speed/
13. M. Zoljoodi, A. Mohseni Arasteh, M. Ghazi Mirsaeid, The effects of internal waves on sound speed in shallow waters of the Persian Gulf. Int. J. Coast. Offshore Eng. **2** (2016). http://ijcoe.org/article-1-53-en.html. http://ijcoe.org/article-1-53-en.pdf

Chapter 3
Opportunistic Transmission in IoUT Networks

3.1 Underwater Communication

3.1.1 Channel Model

Considering a communication link L such as shown in Fig. 3.1, during ϕ_1, between the source node S and the first relay node R_1, the signal-to-noise ratio (SNR) per bit (γ) of an emitted underwater signal at the receiving node can be calculated as [11, 29]

$$\gamma = S_{level} - T_{loss} - N_{level} + D_{index}, \tag{3.1}$$

where S_{level}, T_{loss}, N_{level}, and D_{index} are the source level, transmission loss, noise level, and directivity index, respectively, noting that:

- all quantities are in dB,
- D_{index} is 0 for underwater environments as omnidirectional hydrophones are often used,

and S_{level} is calculated as [11]

$$S_{level} = 10[\log(I) - \log(0.67 \times 10^{-18})], \tag{3.2}$$

where I is the transmitted signal intensity of the sound and can be expressed as [13]

$$I = \frac{p^2}{\rho c} \cos \theta, \tag{3.3}$$

where p is the sound pressure, c is the sound velocity, ρ is the mass density, and θ is the angle between the direction of propagation of the sound and the normal to the surface.

© The Author(s), under exclusive license to Springer Nature Switzerland AG 2021
A. A. Aziz El-Banna, K. Wu, *Machine Learning Modeling for IoUT Networks*,
SpringerBriefs in Computer Science, https://doi.org/10.1007/978-3-030-68567-6_3

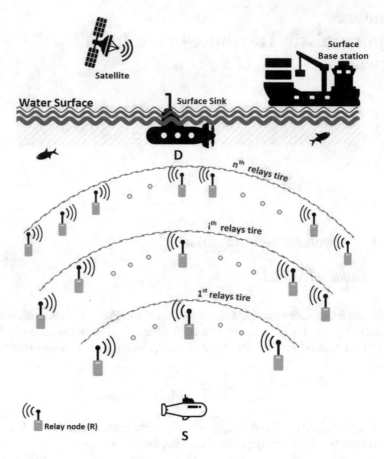

Fig. 3.1 Architecture of a multihop cooperative communication for UWSN

The underwater transmission loss T_{loss} over a distance d (in m) for a signal of frequency f (in kHz) can be calculated as [10]

$$T_{loss} = SS + \alpha(f) \times d \times 10^{-3}, \tag{3.4}$$

where $\alpha(f)$ is the absorption coefficient of sound in seawater in dB/km and can be calculated using Thorp's formula [21], as

$$\alpha(f) = \frac{0.11f^2}{1+f^2} + \frac{44f^2}{4100+f^2} + 2.75 \times 10^{-4}f^2 + 0.003, \tag{3.5}$$

and SS is the spherical spreading that estimates the energy decrease rate and is defined as [10]

$$SS = -10 \log \left(\frac{I_{SS}}{I} \right), \tag{3.6}$$

where I_{SS} is the signal intensity of spherical spreading, and I is the source signal intensity.

The noise level N_{level} is affected by four sources, turbulence, shipping, waves, and thermal noise, and it can be simplified with a practical approximation as [14]

$$N_{level} = 50 - 18 \log f \tag{3.7}$$

Combining the above equations all together, the SNR can be expressed as

$$\gamma = 10 \left[\log(I) - \log(0.67 \times 10^{-18}) \right] - SS - 50 + 18 \log f$$
$$- \left[\left(\frac{0.11 f^2}{1 + f^2} + \frac{44 f^2}{4100 + f^2} + 2.75 \times 10^{-4} f^2 + 0.003 \right) \right.$$
$$\left. \times d \times 10^{-3} \right] \tag{3.8}$$

3.1.2 Transmission System Analysis

A representation for one multihop underwater path between S and D is shown in Fig. 3.2. At the first phase of transmission ϕ_1, S broadcasts its data to their neighbors, and R_1 from the first relay tire decides to participate in the transmission process. Similarly, R_2 from the second relay tire decides to participate and so on till the n-th tire. Each relay node, R_i, can pick out either AF or DF relaying scheme to process and forward the received data stream. Therefore, two homogeneous forwarding schemes are supported where all relays in the multihop path employ the same protocol, i.e., either AF or DF, and hybrid forwarding schemes where relays employ different protocols, e.g., R_1 employs AF and R_2 employs DF or vice versa and so on for other relays.

The received signal at R_1 during ϕ_1 is given by

$$y_{R_1} = h_{SR_1} x + n_{SR_1}, \tag{3.9}$$

Fig. 3.2 A multihop underwater path between S and D

where h_{SR_1} is the channel coefficient from S to R_1, x is the signal transmitted by the S node with transmit power P_s, and n_{SR_1} is the zero-mean additive white Gaussian noise (AWGN) with unit variance at R_1.

Other received signals at each receiving node are determined according to the performed relaying method at the previous transmitting node. In the following, we present the analysis of the received signal at relay R_i in the two cases where its previous relay R_{i-1} performs either AF or DF as follows:

If R_{i-1} performed AF, then during ϕ_i, R_i receives the signal

$$y_{R_i}^{AF} = h_{R_{i-1}R_i} x_a^{(i-1)} + n_{R_{i-1}R_i}, \tag{3.10}$$

where $h_{R_{i-1}R_i}$ is the channel gain coefficient of the ith hop from R_{i-1} to R_i, and $n_{R_{i-1}R_i}$ is the zero-mean AWGN with unit variance at R_i, and the term $x_a^{(i-1)} = \beta_{R_{i-1}} y_{R_{i-1}}$ is the signal transmitted by the R_{i-1} node with transmit power $P_{R_{i-1}}$ after amplified by an amplification factor $\beta_{R_{i-1}}$ which is given by

$$\beta_{R_{i-1}} = \sqrt{\frac{P_{R_{i-1}}}{P_{i-2}|h_{R_{i-2}R_{i-1}}|^2 + 1}}.$$

If R_{i-1} performed DF, then during ϕ_i, R_i receives the signal

$$y_{R_i}^{DF} = h_{R_{i-1}R_i} \hat{x} + n_{R_{i-1}R_i}, \tag{3.11}$$

where $\hat{x}$ is the estimated or decoded signal at R_{i-1}.

At the final stage, the destination node combines the signals come from the different cooperating paths and performs some sort of the conventional diversity combining, e.g., maximum ratio combining or equal gain combining [6].

3.2 Confidence Metric

In this section, we give a brief discussion about the seawater's key physical variables and how they affect the underwater communication. We also show the derivation of the confidence metric (CM) as a selection parameter for the type of transmission scheme, power control, and the employed modulation. Specifically, we investigate these factors: water temperature (T), salinity (S), the water depth (z), the density (ρ), and the internal wave speed (w). Finally, we show how to employ the CM to select the appropriate TPL, the forwarding protocol, and the modulation method.

3.2.1 Derivations of the Confidence Metric

The most applicable transmission for the different underwater environments is performed using acoustic signals [2, 3, 8, 12, 18, 23, 24, 26–29]; therefore, we will firstly analyze the sound speed and show the parameters that affect its propagation. The speed of sound in fluids can be calculated using the Newton–Laplace equation [20]

$$c = \sqrt{\frac{K_s}{\rho}} \tag{3.12}$$

where K_s is the isentropic bulk modulus. Besides, the speed of sound in seawater depends on salinity, temperature, and pressure (hence depth). Note that the pressure p in a liquid is a function of the density and the water depth z as $p = \rho g z$, where g is the acceleration of gravity.

Various models have been derived for the speed of sound in seawater with different accuracy and computational complexities over a wide range of conditions [4, 5, 15, 17]. A simple empirical model in [15] with reasonable accuracy for the world's oceans is employed and is given as follows

$$C(T, S, z) = a_1 + a_2 T + a_3 T^2 + a_4 T^3 + a_5(S - 35)$$
$$+ a_6 z + a_7 z^2 + a_8 T(S - 35) + a_9 T z^3 \tag{3.13}$$

where $a_1 = 1448.96$, $a_2 = 4.591$, $a_3 = 5.304 \times 10^{-2}$, $a_4 = 2.374 \times 10^{-4}$, $a_5 = 1.340$, $a_6 = 1.630 \times 10^{-2}$, $a_7 = 1.675 \times 10^{-7}$, $a_8 = 1.025 \times 10^{-2}$, and $a_9 = -7.139 \times 10^{-13}$.

The variation of sound speed due to internal wave movements is almost equal to the vertical movement of water multiplied by $\frac{d\bar{c}}{dz}$ [32], where $\bar{c}$ is the speed of non-turbulent sound and can be calculated from (3.13). Therefore, an approximation model to estimate the effect of the internal seawater wave on the sound could be stated as

$$C(w) = v \frac{d\bar{c}}{dz} \tag{3.14}$$

where v is the vertical speed of the internal ocean wave and depends on the wavelength λ and the water depth z in shallow enough depths and can be calculated as [1, 16, 22]

$$v = \sqrt{\frac{g\lambda}{2\pi} \tanh\left(2\pi \frac{z}{\lambda}\right)}, \tag{3.15}$$

where in deep water, the hyperbolic tangent approaches 1; hence, the first term in the square root determines the deep water speed. Using limits on the $tanh$ function where $tanh(\Psi) = 1$ for large Ψ and $tanh(\Psi) = \Psi$ for small Ψ, the limiting cases for the velocity expression are obtained as

$$v = \begin{cases} \sqrt{\frac{g\lambda}{2\pi}}, & \text{for deep water,} \quad z > \lambda/2 \\ \sqrt{gz}, & \text{for shallow water,} \quad z < \lambda/2. \end{cases} \quad (3.16)$$

Therefore, the variation of sound speed due to internal wave movements can be estimated as

$$C(w) = \begin{cases} (a_6 + \bar{a}_7 z + \bar{a}_9 T z^2)\sqrt{\frac{g\lambda}{2\pi}}, & \forall z > \lambda/2 \\ (a_6 + \bar{a}_7 z + \bar{a}_9 T z^2)\sqrt{gz}, & \forall z < \lambda/2. \end{cases} \quad (3.17)$$

where $\bar{a}_7 = 2a_7$ and $\bar{a}_9 = 3a_9$.

In consequence, a model that takes into consideration the effect of the KPV parameters T, S, z, ρ, and w on the sound speed can be obtained by combining the significance of Eqs. (3.12), (3.13), and (3.17) together to get the following formula

$$C(T, S, z, \rho, w) = (C(T, S, z) + C(w)) \rho^{-1/2}, \quad (3.18)$$

Moreover, in order to figure out a model describing the spherical spreading, SS term, as a function of the aforementioned KPVs, substitute (3.18) into (3.6) and normalize other parameters to get

$$SS = -10 \log C(T, S, z, \rho, w). \quad (3.19)$$

To define the confidence metric, calculate the symbol and bit error rates, considering QPSK transmission, as [7]

$$P_s \approx 2Q(\sqrt{\gamma_s}), \quad (3.20)$$

$$P_b \approx 2Q(\sqrt{\gamma_b}), \quad (3.21)$$

respectively, where $\gamma_s = 2\gamma_b$ and $Q(\epsilon)$ is defined as the probability that a Gaussian random variable with zero-mean and unit variance exceeds the value ϵ.

The average bit error rate (BER) is defined as [7]

$$P_b = \frac{1}{2}\left[1 - \sqrt{\frac{10^{\frac{\gamma_b}{10}}}{1 + \frac{\gamma_b}{10}}}\right]. \quad (3.22)$$

The confidence metric is defined by the second term in (3.22) and is explicitly given as

$$CM = \sqrt{\frac{10^{\frac{\gamma_b}{10}}}{1 + \frac{\gamma_b}{10}}} \qquad (3.23)$$

where $\gamma = 10 \left[\log(I) - \log(0.67 \times 10^{-18}) \right] + 10 \log C(T, S, z, \rho, w) - 50 +$
$18 \log f - \left[\left(\frac{0.11 f^2}{1 + f^2} + \frac{44 f^2}{4100 + f^2} + 2.75 \times 10^{-4} f^2 + 0.003 \right) \times d \times 10^{-3} \right].$

3.2.2 TPL Control and Adaptive Modulation

The transmitted signal intensity, defined in (3.3), for a spherical sound wave can be determined as a function of the sound power at a distance r from the center of the sphere by using the inverse square law as [11, 13]

$$I = \frac{P_{sound}}{A(r)}, \qquad (3.24)$$

where P_{sound} is the sound power and $A(r) = 4\pi r^2$ is the surface area of a sphere with radius r. The transmitted signal intensity can be adjusted by controlling the sound power output from the power amplifier circuit implemented in the transmitter noting that various available commercial WSN motes usually have many different transmission power levels that we can choose the proper one among them, for example, the Mica2 motes, the most employed mote for the experimental WSN research [31], have 26 different power levels with different power consumption for transmission with output antenna power values extending from -20 dBm to 5 dBm.

Furthermore, the commonly used modulation methods in acoustic communication are the phase shift keying (PSK) variations (e.g., binary BPSK, quadrature QPSK, and 8PSK) besides the quadrature amplitude modulation (QAM) methods [25, 30] where MPSK is considered the most commonly used modulation method in underwater transmission. The basic idea of the PSK is changing the phase of the modulating signal according to the information signal while fixing its amplitude and frequency. The MPSK signal can be represented by

$$x_{MPSK}(t) = \cos(w_c t + \theta_m), \qquad (3.25)$$

where A represents the amplitude, w_c represents the angular frequency, and phase θ_m is represented by a uniformly spaced set of phase angles

$$\theta_m = \frac{2(m-1)\pi}{M}, \, m = 1, 2, \ldots, M, \qquad (3.26)$$

where M represents the number of symbols and the phase interval between two adjacent signals in the modulation signal is $2\pi/M$. The modulation method can be adjusted by changing the phase θ_m, i.e., by modifying the number of symbols in (3.26).

3.2.3 Employing the Confidence Metric in the Transmission Framework

In the following, we illustrate how to employ the confidence metric to make use of the good channel state in an opportunistic way to determine the proper candidates of the transmission framework. Firstly, to choose the appropriate relaying scheme among AF and DF, we introduce a system predefined parameter, ξ, which represents the threshold that the CM will be compared with to select the suitable forwarding protocol. The threshold ξ acts as the tradeoff between the network performance and the computational complexity at the relay nodes which is another perspective of the energy consumption. With a large threshold, we get a small error rate, i.e., high performance is achieved but more computations are required and hence more energy is consumed and vice versa.

Therefore, based on the computed value of CM, we can adaptively choose the proper type between AF and DF to use it in the relay node as follows. When the underwater channel is good, it will incur fewer errors in the transmitted signal, i.e., the SNR is high and the confidence metric will be high as well, and then we choose AF relaying because in that way we guarantee error-free forwarded signals to the destination node, but if DF is used in this case, it will demand much higher computational complexity than AF and accordingly consume more energy without considerable merits. In other words, we take advantage of the property of the AF scheme where its error performance is almost the same as that of the DF scheme insofar as the channel condition is good [9]. On the other side, when the channel conditions become bad, CM will be low and DF becomes the proper choice because AF will perform badly in poor channels as it will propagate signals with noise and interference. In this way, we guarantee that the selected scheme is the one that dynamically compromises between the network performance and the computational complexity and as a result, ensures reliable low-power networking.

Moreover, depending on the CM, we select a certain modulation method by changing the number M in (3.25) to move from one PSK method to another one, c.g., from BPSK to QPSK to 8PSK or vice versa, and hence depending on the channel state, we could select the appropriate method that achieves the best compromise between the noise robustness and the achieved data rate. In this part, we employ two methods, i.e., BPSK for noisy channels and QPSK for good channels, and we use the predefined threshold ξ to distinguish between these two states.

In the same manner, knowing the CM, we further select a certain power level from the available levels defined by the mote, and according to it, we change the

---

Algorithm Opportunistic transmission scheme based on seawater's key physical variables

---

1: Get $T, S, z, \rho, w, d, f, P_{sound}$
2: Compute $C(T, S, z, \rho, w)$ from (18)
3: Compute $SS(T, S, z, \rho, w)$ from (19)
4: Compute $\gamma_b(T, S, z, \rho, w, I, f, d)$ from (8) using (19)
5: Compute $CM(\gamma_b)$ from (23)
6: **if** $CM(\gamma_b) > \xi$ **then**
7:     use AF relaying scheme
8:     use QPSK modulation
9: **else**
10:     use DF relaying scheme
11:     use BPSK modulation
12: **end if**
13: **if** $(d_{RRx} > \varrho)$ or $(CM(\gamma_b) < \xi)$ **then**
14:     **if** $TPL < TPL_{max}$ **then**
15:         use the next TPL
16:     **else**
17:         use the current TPL_{max}
18:     **end if**
19: **else if** $(d_{RRx} < \varrho)$ or $(CM(\gamma_b) > \xi)$ **then**
20:     **if** $TPL > TPL_{min}$ **then**
21:         use the previous TPL
22:     **else**
23:         use the current TPL_{min}
24:     **end if**
25: **end if**

Fig. 3.3 Algorithm 3.1

output power P_{sound} that changes the sound signal intensity. Therefore, we introduce a relay predefined parameter, ϱ, which represents the threshold that will be checked against the distance between the transmitting node and the receiving one (denoted as d_{RRx}), meaning that, if the distance between the nodes is far and greater than ϱ, the relay increases its transmission power by selecting a higher TPL as long as the current TPL is still below the maximum allowed value TPL_{max}, and as a result, the transmitted signal intensity increases as well, as shown in (3.24), to compensate for the expected transmission loss due to large distances of the communication link and vice versa, i.e., if the distance in between nodes is small, the relay expects low transmission loss in the channel and decreases the sound power P_{sound} by tuning to a lower TPL as long as the current TPL is still above the minimum allowed value TPL_{min} and consequently saves the consumed transmission power emphasizing reliable and low-power transmission.

The opportunistic transmission scheme and the signal forwarding process are summarized in the algorithm (Fig. 3.3).

3.3   Performance Analysis

Considering the ocean profile dataset in [19] collected from more than 20,000
separate archived datasets contributed by institutions, project, government agencies,
and individual investigators from the United States and around the world, we have
numerous sets of measurements of ocean KPVs vs. depth, e.g., temperature, salinity,
and wind speed. The KPV numerical data, shown in Fig. 3.4, is used for computing
the γ quantities, and a 1K samples of γ values is shown in Fig. 3.5. These values are
then analyzed to compute the CM using (3.23) and to estimate the threshold value
ξ that is used to indicate which relaying scheme and modulation methods should
be used as will be illustrated later at the end of this section. Additionally, the effect
of the speed of sound by the factors T, S, z, ρ, and w is demonstrated in Fig. 3.6
to highlight the direct impact of the selected parameters on the sound speed and
accordingly on the transmitted signal intensity in the underwater environment.

For the sake of simulation of the network depicted in Fig. 3.1 and without loss of
generality, we considered three paths (in the same manner as the two paths shown
in Fig. 1.1 in Chap. 1) from the source to destination, namely, L_1, L_2, and L_3, and
assumed that the path L_1 employed three relays R_{11}, R_{12}, and R_{13} , while the paths
L_2 and L_3 employed two relays R_{21}, R_{22}, and R_{31}, R_{32}, respectively. For the path

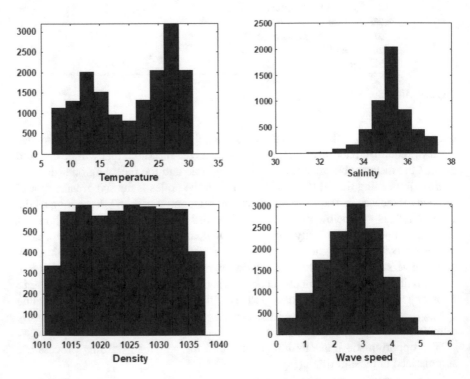

Fig. 3.4 Dataset distributions of the temperature, salinity, density, and wave speed

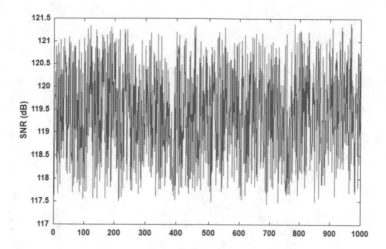

Fig. 3.5 Plot of 1K samples of γ values

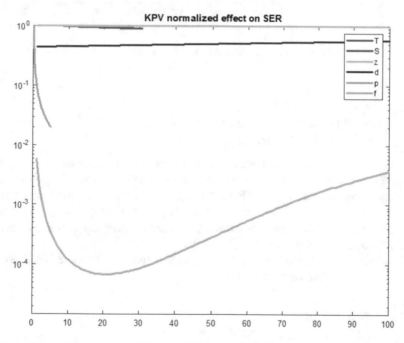

Fig. 3.6 Effect of temperature, salinity, depth, distances between nodes, transmitting power, and acoustic signal frequency on the SER

L_1, there exist eight possible combinations to transmit the signals between S and D via this path based on the employed relaying scheme at the relay nodes. The eight combinations are:

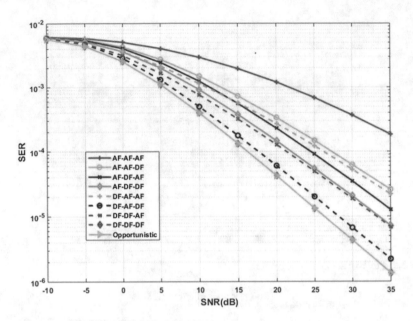

Fig. 3.7 SER performance of the different transmission schemes at $\xi = \mu$ for the path with three relays

1. G_1 : AF-AF-AF and DF-DF-DF,
2. G_2 : AF-AF-DF, AF-DF-AF, and DF-AF-DF,
3. G_3 : AF-DF-DF, DF-AF-DF, and DF-DF-AF.

Furthermore, we partition these eight combinations into three groups G_1 : G_3, where G_1 represents the homogeneous forwarding schemes where all relays employ the same protocol, i.e., either AF or DF, G_2 represents the hybrid forwarding schemes where any two relays out of three employ the AF protocol, and finally G_3 represents the other hybrid forwarding schemes where any two relays out of three employ the DF protocol.

However, for the paths L_2 and L_3, there exist four possible combinations for transmission. The four combinations are:

1. G_1 : AF-AF,
2. G_2 : AF-DF and DF-AF,
3. G_3 : DF-DF.

Monte Carlo simulations are carried out to examine the performance of the proposed opportunistic cooperative scheme against the conventional AF and DF schemes at threshold ξ equals to the mean μ of the CM, as shown in Fig. 3.7. The total distance between the S and D nodes is divided equally between the relay nodes, i.e., we assume that the distance between any communicating pair equals the quarter

Table 3.1 Simulation parameters

Parameter	Value
Number of QPSK symbols	1×10^6
Total transmitted power for S and R nodes (W)	2
Power split factor between S and R nodes	0.7
Channel model	Rayleigh
Path loss exponent	3
Temperature range (°C)	0–30
Salinity range (ppt)	20–37
Density range (kg/m^3)	1000–1050
Wave speed range (m/s)	1.5–4
Water depth range (m)	0–3000
Transmitter power (W)	1
Transmitter range (m)	300
Transmission distance (m)	300
Frequency (K Hz)	10

of the total distance, and we assume perfect channel knowledge at each receiving node and the simulation parameters are given in Table 3.1.

For the path that employs three relays, i.e., path L_1, the symbol error rate (SER) performance of the proposed opportunistic transmission scheme is comparable to the highest performance DF-DF-DF scheme, i.e., the scheme which employs DF at all the relay nodes, and outperforms all other hybrid schemes and the homogeneous AF-AF-AF. To distinguish the performance differences between employing the AF and DF forwarding protocols at the relay nodes, Figs. 3.8, 3.9, and 3.10 present the individual performance of each one of the above divided groups $G_1 : G_3$. Figure 3.8 compares between the two homogeneous forwarding schemes AF-AF-AF and DF-DF-DF, and as shown the gap between employing the AF scheme at all relays versus employing the DF is huge and equals 15 dB at $SER = 10^{-3}$. Figure 3.9 shows the performance of the group of forwarding schemes when two relays apply AF, the performances are comparable and depends on the SNR as well as the order of the applied schemes e.g. when the first two relays employ AF, this gives the lower performance. Figure 3.10 shows the performance of the group of forwarding schemes; when two relays apply DF, the schemes performances are different, and the best performance is achieved when the AF is performed in the middle relay.

For the paths that implements two relays, i.e., L_2 and L_3, a similar analysis is given in Fig. 3.11, and the SER performance of the proposed opportunistic transmission scheme is also comparable to the highest performance DF-DF scheme and outperforms all other hybrid schemes and the homogeneous AF-AF.

However, the proposed scheme requires much less computational complexity than the fully DF scheme, i.e., DF-DF-DF in the three-relay path and DF-DF in the two-relay paths. For example, while calculating the performance of the opportunistic schemes for a total samples of 10^4 sample, Algorithm (Fig. 3.3) picked up the AF and DF at each relay in each scheme of the eight possible schemes as calculated in Table 3.2. To roughly determine the percentage of complexity savings

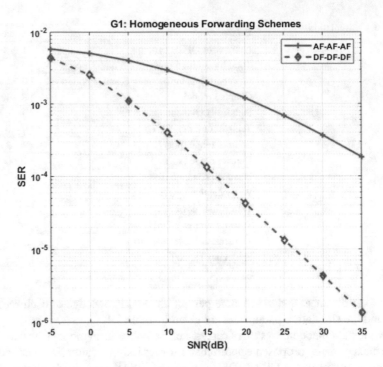

Fig. 3.8 SER performance of the transmission group G1: homogeneous forwarding schemes

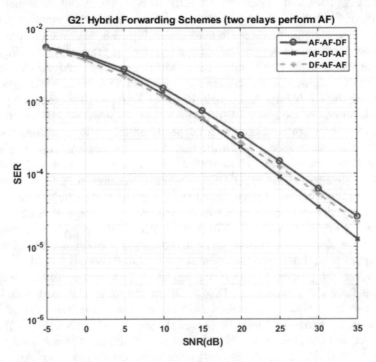

Fig. 3.9 SER performance of the transmission group G2: hybrid forwarding schemes (two relays perform AF)

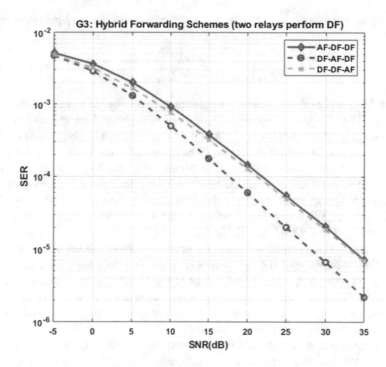

Fig. 3.10 SER performance of the transmission group G3: hybrid forwarding schemes (two relays out of three perform DF)

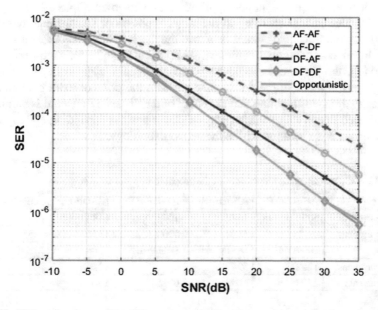

Fig. 3.11 SER performance of the different transmission schemes at $\xi = \mu$ for the paths with two relays

in using the complex fully DF scheme, we introduce an efficiency parameter as

$$eff = \frac{\sum_{schm=1}^{8} AF_{sel}}{\sum_{schm=1}^{8} AF_{sel} + \sum_{schm=1}^{8} DF_{sel}} \%, \tag{3.27}$$

where AF_{sel} and DF_{sel} represent the accumulative number of times that the scheme selects AF and DF is the method of choice, respectively, and are summed on all the selected eight schemes $schm = 1 : 8$ of the three aforementioned groups as shown in Table 3.2. The % eff for the data showed in Table 3.2 is 48.53%. Similar data are obtained when checking to distinguish between QPSK and BPSK, where BPSK was selected in a total number of 13,131 to achieve the proper noise immunity when the channel is not good, i.e., $CM(\gamma_b) > \xi$, while the higher data rate modulation QPSK was selected 12,381 times. The threshold values ξ and ϱ have been set heuristically as following: ξ is set mid-way in the CM range, i.e., $\xi = \mu$ and hence any value of CM above the CM mean will drive the proposed scheme to select AF and QPSK depending on the estimated good channel state which allows noise immune and high data rate transmission, while ϱ was chosen as the third of the total distance between the node and the destination, i.e., when d_{RRx} is greater than the third of the total distance, we should increase the power by activating the higher TPL to ensure good reception at the receiving node.

In that manner, we ensure that the proposed opportunistic scheme consumes less energy in transmission, and as a result, it guarantees longer lifetime for battery-based underwater sensor nodes.

Finally, an empirical analysis was conducted to estimate the proper threshold values depending on the target application as shown in Fig. 3.12. The proper threshold value is set according to the target application, i.e., for energy-efficient networking, the threshold should be above the mean value to significantly depend on the lower computational complexity AF scheme, while for high-performance networking, the threshold should be below the mean value to depend more on the high performance of the DF scheme but with the price of extra computational complexity. It is worth noting that the effect of the distance threshold is similar and is considered linear with the transmission power besides depending entirely on the instantaneous node location within the network topology.

Table 3.2 Scheme selection recurrences in the proposed opportunistic scheme

Scheme	AF-AF-AF	AF-AF-DF	AF-DF-AF	AF-DF-DF	DF-AF-AF
AF recurrences	6033	1072	1072	648	1710
DF recurrences	0	536	536	1296	855
Scheme	DF-AF-DF	DF-DF-AF	DF-DF-DF	Total	
AF recurrences	1252	594	0	12,381	
DF recurrences	2504	1188	6216	13,131	

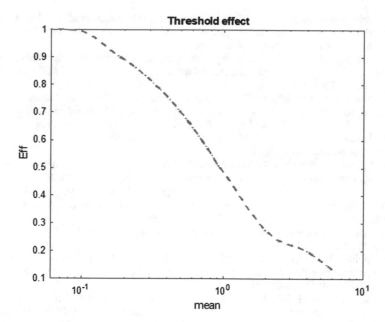

Fig. 3.12 SER performance of the different transmission schemes at $\xi = \mu$ for the paths with two relays

References

1. W. Bascom, *Waves and Beaches* (Doubleday, New York, 1964)
2. Y. Chen, Z. Wang, L. Wan, H. Zhou, S. Zhou, X. Xu, OFDM-modulated dynamic coded cooperation in underwater acoustic channels. IEEE J. Ocean. Eng. **40**(1), 159–168 (2015). https://doi.org/10.1109/JOE.2014.2304254
3. R.W.L. Coutinho, A. Boukerche, L.F.M. Vieira, A.A.F. Loureiro, Underwater wireless sensor networks: A new challenge for topology control-based systems. ACM Comput. Surv. **51**(1), 19:1–19:36 (2018). https://doi.org/10.1145/3154834
4. V.A. Del Grosso, New equation for the speed of sound in natural waters (with comparisons to other equations). J. Acoust. Soc. Am. **56**(4), 1084–1091 (1974). https://doi.org/10.1121/1.1903388
5. B.D. Dushaw, P.F. Worcester, B.D. Cornuelle, B.M. Howe, On equations for the speed of sound in seawater. J. Acoust. Soc. Am. **93**(1), 255–275 (1993). https://doi.org/10.1121/1.405660
6. T. Eng, N. Kong, L.B. Milstein, Comparison of diversity combining techniques for Rayleigh-fading channels. IEEE Trans. Commun. **44**(9), 1117–1129 (1996)
7. A. Goldsmith, *Wireless Communications* (Cambridge University Press, Cambridge, 2005). https://doi.org/10.1017/CBO9780511841224
8. J. Heidemann, Y. Wei, J. Wills, A. Syed, L. Yuan, Research challenges and applications for underwater sensor networking, in *IEEE Wireless Communications and Networking Conference, 2006. WCNC 2006*, vol. 1 (2006), pp. 228–235
9. J. Jeon, Y. Shim, H. Park, Optimal power allocation with hybrid relaying based on the channel condition. Appl. Sci. **8**(5) (2018). https://doi.org/10.3390/app8050690. http://www.mdpi.com/2076-3417/8/5/690

10. K.A. Johannesson, R.B. Mitson, Food and Agriculture Organization of the United Nations, *Fisheries Acoustics - a Practical Manual for Aquatic Biomass Estimation* (Food and Agriculture Organization of the United Nations, Rome, 1983)
11. C.C. Kao, Y.S. Lin, G.D. Wu, C.J. Huang, A comprehensive study on the internet of underwater things: Applications, challenges, and channel models. Sensors **17**(7), 1477 (2017). https://doi.org/10.3390/s17071477
12. A. Khan, I. Ali, A.U. Rahman, M. Imran, H. Mahmood, Co-EEORS: Cooperative energy efficient optimal relay selection protocol for underwater wireless sensor networks. IEEE Access **6**, 28777–28789 (2018). https://doi.org/10.1109/ACCESS.2018.2837108
13. L.D. Landau, E. Lifshitz, *Fluid Mechanics: Course of Theoretical Physics*, vol. 6, 2nd edn. (Butterworth-Heinemann, Oxford, 2010)
14. D.E. Lucani, M. Stojanovic, M. Medard, On the relationship between transmission power and capacity of an underwater acoustic communication channel, in *OCEANS 2008 - MTS/IEEE Kobe Techno-Ocean* (2008), pp. 1–6. https://doi.org/10.1109/OCEANSKOBE.2008.4531073
15. K.V. Mackenzie, Discussion of sea water sound–speed determinations. J. Acoust. Soc. Am. **70**(3), 801–806 (1981). https://doi.org/10.1121/1.386919
16. N. Mayo, Ocean waves-their energy and power. Phys. Teacher **35**(6), 352–356 (1997). https://doi.org/10.1119/1.2344718
17. C.S. Meinen, D.R. Watts, Further evidence that the sound-speed algorithm of del grosso is more accurate than that of Chen and Millero. J. Acoust. Soc. Am. **102**(4), 2058–2062 (1997). https://doi.org/10.1121/1.419655
18. H. Nasir, N. Javaid, H. Ashraf, S. Manzoor, Z.A. Khan, U. Qasim, M. Sher, CoDBR: Cooperative depth based routing for underwater wireless sensor networks, in *2014 Ninth International Conference on Broadband and Wireless Computing, Communication and Applications* (2014), pp. 52–57. https://doi.org/10.1109/BWCCA.2014.45
19. NCEI standard product: World ocean database (WOD). https://catalog.data.gov/dataset/ncei-standard-product-world-ocean-database-wod
20. The Newton–Laplace equation and speed of sound (2015). https://www.thermaxxjackets.com/newton-laplace-equation-sound-velocity/. Accessed 15 Jul 2019
21. R.J. Urick, *Principles of Underwater Sound*, 3rd edn. (Peninsula Pub, Los Altos Hills, 1996)
22. Velocity of idealized ocean waves. http://hyperphysics.phy-astr.gsu.edu/hbase/watwav.html
23. P. Wang, X. Zhang, Energy-efficient relay selection for QoS provisioning in MIMO-based underwater acoustic cooperative wireless sensor networks, in *2013 47th Annual Conference on Information Sciences and Systems (CISS)* (2013), pp. 1–6. https://doi.org/10.1109/CISS.2013.6624269
24. P. Wang, L. Zhang, V.O.K. Li, Asynchronous cooperative transmission for three-dimensional underwater acoustic networks. IET Commun. **7**(4), 286–294 (2013). https://doi.org/10.1049/iet-com.2012.0314
25. Y. Wang, H. Zhang, Z. Sang, L. Xu, C. Cao, T.A. Gulliver, Modulation classification of underwater communication with deep learning network. Comput. Intell. Neurosci. 2019, Article ID 8039632, 12 (2019). https://doi.org/10.1155/2019/8039632
26. P. Xie, J.H. Cui, L. Lao, VBF: Vector-based forwarding protocol for underwater sensor networks, in *NETWORKING 2006. Networking Technologies, Services, and Protocols; Performance of Computer and Communication Networks; Mobile and Wireless Communications Systems*, ed. by F. Boavida, T. Plagemann, B. Stiller, C. Westphal, E. Monteiro (Springer Berlin Heidelberg, Berlin, Heidelberg, 2006), pp. 1216–1221
27. F. Xu, L. Yang, Two-way relay underwater acoustic communications with multiuser decision-feedback detection and relay preprocessing, in *2012 1st IEEE International Conference on Communications in China (ICCC)* (2012), pp. 602–607. https://doi.org/10.1109/ICCChina.2012.6356955
28. H. Yan, Z.J. Shi, J.H. Cui, D B R: Depth-based routing for underwater sensor networks, in *NETWORKING 2008 Ad Hoc and Sensor Networks, Wireless Networks, Next Generation Internet*, ed. by A. Das, H.K. Pung, F.B.S. Lee, L.W.C. Wong (Springer Berlin Heidelberg, Berlin, Heidelberg, 2008), pp. 72–86

29. H. Yang, B. Liu, F. Ren, H. Wen, C. Lin, Optimization of energy efficient transmission in underwater sensor networks, in *GLOBECOM 2009 - 2009 IEEE Global Telecommunications Conference* (2009), pp. 1–6. https://doi.org/10.1109/GLOCOM.2009.5425484
30. Z. Yang, Y.R. Zheng, Iterative channel estimation and turbo equalization for multiple-input multiple-output underwater acoustic communications. IEEE J. Ocean. Eng. **41**(1), 232–242 (2016). https://doi.org/10.1109/JOE.2015.2398731
31. H.U. Yildiz, B. Tavli, H. Yanikomeroglu, Transmission power control for link-level handshaking in wireless sensor networks. IEEE Sensors J. **16**(2), 561–576 (2016). https://doi.org/10.1109/JSEN.2015.2486960
32. M. Zoljoodi, A. Mohseni Arasteh, M. Ghazi Mirsaeid, The effects of internal waves on sound speed in shallow waters of the Persian Gulf. Int. J. Coast. Offshore Eng. **2** (2016). http://ijcoe.org/article-1-53-en.html. http://ijcoe.org/article-1-53-en.pdf

Chapter 4
Localization and Positioning for Underwater Networks

4.1 Introduction

In this chapter, we address the problem of localization and positioning in underwater acoustic networks considering the surrounding environmental effects at that place. Most of the proposed underwater positioning algorithms are using a constant predetermined speed of sound; however, the underwater speed of sound fluctuates with the change of different physical seawater properties, e.g., temperature, salinity, and depth or pressure [2, 10, 14]. Furthermore, few studies considered the effect of variable speed on localization, e.g., [3, 10, 14]; however, they limit their studies to the effect of few physical variables, i.e., temperature, salinity, and depth, depending on the simplified empirical Mackenzie model which only considers these three variables. However, further studies proved that other factors still affect the speed of sound significantly, e.g., density, conductivity, and internal waves effect [11]. Moreover, in [10], the proposed algorithm required knowing the locations of four nodes in the start-up phase which may not be available in many scenarios.

4.2 System Modeling Using TDoA

We consider the underwater position determination system as shown in Fig. 4.1, where three buoys are assumed to know their locations via GPS system, and these buoys are considered anchors. An underwater node or a remotely operated underwater vehicle (ROV) will use the buoys location information in determining its own position. The unknown position of the ROV is denoted (x_r, y_r), the position of anchor i is (x_i, y_i), and r_i is the Euclidean distance between the ROV and anchor $i, \forall i \in 1, 2, 3$.

Using the trilateration approach to compute the position estimate, we have the following common estimation methods to determine the distances r_i: received

© The Author(s), under exclusive license to Springer Nature Switzerland AG 2021
A. A. Aziz El-Banna, K. Wu, *Machine Learning Modeling for IoUT Networks*,
SpringerBriefs in Computer Science, https://doi.org/10.1007/978-3-030-68567-6_4

Fig. 4.1 An underwater position determination system based on TDoA

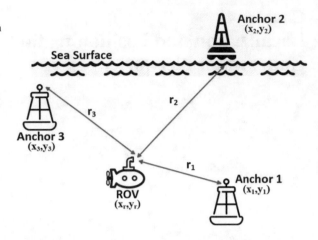

signal strength indicator (RSSI), angle of arrival (AoA), time of arrival (ToA), and time difference of arrival (TDoA) [6, 15]. Considering the TDoA technique, and for calculation simplicity without loss of generality, we consider anchor 1 as the closest anchor to the ROV and has the earliest time of arrival; therefore, the position of anchor 1 is set as the origin of the Cartesian coordinate system, i.e., $(x_1, y_1) = (0, 0)$. The unknown position (x_r, y_r) can be estimated by using the following set of equations [13]

$$r_1^2 = x_r^2 + y_r^2 \tag{4.1}$$

$$r_2^2 = (x_2 - x_r)^2 + (y_2 - y_r)^2 \tag{4.2}$$

$$r_3^2 = (x_3 - x_r)^2 + (y_3 - y_r)^2. \tag{4.3}$$

Defining the time difference associated with anchor i as $(t_i - t_1)$ and defining the distance differences r_{i1} as [13]

$$r_{i1} \triangleq r_i - r_1 = (t_i - t_1)v, \tag{4.4}$$

where v is the speed of ultrasound waves propagated in water. Substitute (4.4) into (4.2), (4.3), we get

$$- x_2 x_r - y_2 y_r = r_{21} r_1 + \frac{1}{2}(r_{21}^2 - K_2^2) \tag{4.5}$$

$$-x_3x_r - y_3y_r = r_{31}r_1 + \frac{1}{2}(r_{31}^2 - K_3^2) \tag{4.6}$$

where $K_i^2 = x_i^2 + y_i^2$. Rewriting in matrix form gives

$$\mathbf{H}\mathbf{x} = r_1\mathbf{c} + \mathbf{d}, \tag{4.7}$$

where

$$\mathbf{H} = \begin{bmatrix} x_2 & y_2 \\ x_3 & y_3 \end{bmatrix}, \mathbf{x} = \begin{bmatrix} x_r \\ y_r \end{bmatrix}, \mathbf{c} = \begin{bmatrix} -r_{21} \\ -r_{31} \end{bmatrix}, \mathbf{d} = \frac{1}{2} \begin{bmatrix} K_2^2 - r_{21}^2 \\ K_3^2 - r_{31}^2 \end{bmatrix}.$$

To solve for $\mathbf{x}$, rearrange (4.7) to get

$$\mathbf{x} = r_1\mathbf{H}^{-1}\mathbf{c} + \mathbf{H}^{-1}\mathbf{d}. \tag{4.8}$$

Substituting (4.8) into (4.13) gives a quadratic equation in r_1. To obtain the final solution for $\mathbf{x}$, the quadratic equation is solved for r_1 and substitute the positive root back into (4.8).

4.3   TDoA-Based Positioning Approach

The speed of sound underwater v in literature is assumed to be constant and equals 1500 m/s, while few research consider the effect of the KPV parameters on the propagation of the sound waves in water, e.g., [2, 4, 10, 11], but they build on the simplified Mackenzie model [8] that considers only the effect of temperature T, salinity S, and water depth z on the sound speed and calculated it as

$$v(T, S, z) = a_1 + a_2T + a_3T^2 + a_4T^3 + a_5(S - 35)$$
$$+ a_6z + a_7z^2 + a_8T(S - 35) + a_9Tz^3 \tag{4.9}$$

where $a_1 = 1448.96$, $a_2 = 4.591$, $a_3 = 5.304 \times 10^{-2}$, $a_4 = 2.374 \times 10^{-4}$, $a_5 = 1.340$, $a_6 = 1.630 \times 10^{-2}$, $a_7 = 1.675 \times 10^{-7}$, $a_8 = 1.025 \times 10^{-2}$, and $a_9 = -7.139 \times 10^{-13}$.

However, the speed of ultrasound waves in water v still changes with various physical properties such as the in situ density ρ [11] and the internal waves effect w [9]. Water density is considered the most dominant thermodynamic property of seawater for oceanic circulation studies [11], and it depends on heat content and salinity and varies slightly with pressure; therefore, it has a significant effect on the speed of sound. The density has an inverse relationship with the speed of sound as defined by the Newton–Laplace equation [16]

$$v(\rho) = \sqrt{\frac{K_s}{\rho}}. \tag{4.10}$$

where K_s is the isentropic bulk modulus.

On the other hand, the seawater internal waves are considered one of the most important factors that make temporal shifts of the seawater physical properties [17]. The variations of internal waves produce a nonuniform distribution of the refractive index of seawater, and that causes fluctuations for the ultrasonic sound propagation in the water [9]. The effect of the internal waves w on the sound speed can be expressed as [9]

$$v(w) = v_o \left(1 + \frac{1}{\sqrt{2\pi\sigma^2}} exp\left(-\frac{z^2}{2\sigma^2} \right) \right) \tag{4.11}$$

where v_o is the speed of the ultrasonic sound in the absence of internal waves and σ is the standard deviation.

In consequence, a model that takes into consideration the effect of all the studied KPV parameters, i.e., T, S, z, ρ, and w, on the sound speed can be obtained from combining Eqs. (4.9), (4.10), and (4.11) to get the following formula

$$v(T, S, z, \rho, w) = v(T, S, z)\sqrt{\frac{K_s}{\rho}} \left(1 + \frac{1}{\sqrt{2\pi\sigma^2}} exp\left(-\frac{z^2}{2\sigma^2} \right) \right) \tag{4.12}$$

Using (4.12), we could determine accurate positions of the required nodes. The proposed positioning scheme is summarized in Algorithm 4.1 shown in Fig. 4.2.

4.4   System Modeling Using RSS Method

We consider the underwater position determination system defined in Fig. 4.3, where two buoys are assumed to know their locations via GPS system; these buoys are considered anchors that the submarine or ROV will use their location information in determining its own position. The relative position of the ROV can be calculated as [1, 14]

$$x = \frac{(r_1^2 - r_2^2 + d^2)}{2d},$$

$$y = \sqrt{r_1^2 - x^2 - h^2}, \tag{4.13}$$

Algorithm   Underwater positioning determination approach

1: **Inputs:** T, S, z, ρ, w, t_i
2: **Output:** (x_r, y_r)
    Initialization phase:
3: Compute the sound speed $v(T, S, z, \rho, w)$ at that place:
4: $v(T, S, z) = a_1 + a_2 T + a_3 T^2 + a_4 T^3 + a_5 (S - 35) + a_6 z + a_7 z^2 + a_8 T(S - 35) + a_9 T z^3$
5: $v(T, S, z, \rho, w) = v(T, S, z) \sqrt{\dfrac{K_s}{\rho}} \left(1 + \dfrac{1}{\sqrt{2\pi\sigma^2}} exp\left(-\dfrac{z^2}{2\sigma^2}\right)\right)$
    Distance Estimation phase:
6: Compute $r_{i1} = (t_i - t_1)v, \quad \forall i \in 2, 3$
7: Substitute $\mathbf{x} = r_1 \mathbf{H}^{-1}\mathbf{c} + \mathbf{H}^{-1}\mathbf{d}$ into $r_1^2 = x_r^2 + y_r^2$
8: select the positive root for r_1
    Position Estimation phase:
9: substitute r_1 into $\mathbf{x} = r_1 \mathbf{H}^{-1}\mathbf{c} + \mathbf{H}^{-1}\mathbf{d}$
10: return $\mathbf{x} = [x_r; y_r]$
    Refinement phase:
11:    Iterate to fine tune the location
12: **End**

Fig. 4.2 Algorithm 4.1

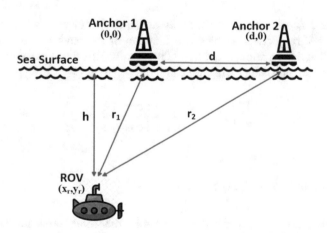

Fig. 4.3 An underwater position determination system based on RSS

where h is the depth of the underwater node and can be found using the hydrostatic pressure p information as $h \approx 10p$, d is the distance between the two anchors, and r_i is the Euclidean distance between the ROV and anchor i, $\forall i \in 1, 2$, and anchor 1 is assumed the origin of the system, i.e., its (x, y) location is $(0, 0)$.

Using RSSI, r_i can be determined using the regression model defined in [12] as

$$r_i = \frac{RSS - 12}{35} \tag{4.14}$$

where RSS is the received signal strength in dB. The RSS is a function of the ratio between the transmitted power P_{tx} and the received power P_{rx} [6]. In underwater communication, the carrier is the sound, therefore we can define the signal intensity in terms of the sound pressure p and the sound velocity v as [5, 7]

$$I = \frac{p^2}{\rho v} \cos \theta, \tag{4.15}$$

where ρ is the mass density and θ is the angle between the direction of propagation of the sound and the normal to the surface.

Equation (4.14) can be written in terms of the sound intensity as

$$r_i = \frac{I_{tx} - I_{rx} - 12}{35} \tag{4.16}$$

where I_{tx} and I_{rx} are the sound intensity at the places of the transmitter and the receiver nodes, respectively, which is a function of the physical properties of the environment at that location, i.e., different pressures, densities, temperatures, speeds of sound, and so on. Using (4.15), we can write (4.16) as

$$r_i = \frac{1}{35} \left(10 \log \left(\frac{\rho_2 v_2 p_1^2 \cos \theta_1}{\rho_1 v_1 p_2^2 \cos \theta_2} \right) - 12 \right). \tag{4.17}$$

4.5   RSS Positioning Approach

Following the analysis presented in Sect. 4.3, the same procedure is applied. Therefore, the resulting positioning scheme for the RSS method is summarized in Algorithm 4.2 shown in Fig. 4.4.

Algorithm Underwater Positioning determination approach

1: **Inputs:** T, S, z, d, ρ, w, t_i

2: **Output:** (x_r, y_r)

  Initialization phase:

3: Compute the sound speed $v(T, S, z, \rho, w)$ at that place:

4: $v(T, S, z) = a_1 + a_2 T + a_3 T^2 + a_4 T^3 + a_5(S - 35) + a_6 z + a_7 z^2 + a_8 T(S - 35) + a_9 T z^3$

5: $v(T, S, z, \rho, w) = v(T, S, z) \sqrt{\dfrac{K_s}{\rho}} \left(1 + \dfrac{1}{\sqrt{2\pi\sigma^2}} exp\left(-\dfrac{z^2}{2\sigma^2} \right) \right)$

  Distance Estimation phase:

6: Compute $r_i = \dfrac{I_{tx} - I_{rx} - 12}{35}$

  Position Estimation phase:

7: compute $x_r = \dfrac{(r_1^2 - r_2^2 + d^2)}{2d}, y_r = \sqrt{r_1^2 - x^2 - h^2},$

8: return $[x_r; y_r]$

  Refinement phase:

9: Iterate to fine tune the location

10: **End**

Fig. 4.4 Algorithm 4.2

References

1. U. Bekcibasi, M. Tenruh, Increasing RSSI localization accuracy with distance reference anchor in wireless sensor networks. Acta Polytech. Hung. **11**(8), 103–120 (2014)
2. A. Caiti, E. Crisostomi, A. Munafò, Physical characterization of acoustic communication channel properties in underwater mobile sensor networks, in *Sensor Systems and Software* (Springer, Berlin, 2009), pp. 111–126
3. J.E. Garcia, Positioning of sensors in underwater acoustic networks, in *Proceedings of OCEANS 2005 MTS/IEEE*, vol. 3 (2005), pp. 2088–2092. https://doi.org/10.1109/OCEANS.2005.1640068
4. J.E. Garcia, Adapted distributed localization of sensors in underwater acoustic networks, in *OCEANS 2006 - Asia Pacific* (2006), pp. 1–4. https://doi.org/10.1109/OCEANSAP.2006.4393815
5. C.C. Kao, Y.S. Lin, G.D. Wu, C.J. Huang, A comprehensive study on the internet of underwater things: Applications, challenges, and channel models. Sensors **17**(7), 1477 (2017). https://doi.org/10.3390/s17071477
6. H. Karl, A. Willig, *Protocols and Architectures for Wireless Sensor Networks* (Wiley, Weinheim, 2005)
7. L.D. Landau, E. Lifshitz, *Fluid Mechanics: Course of Theoretical Physics*, vol. 6, 2nd edn. (Butterworth-Heinemann, Oxford, 2010)
8. K.V. Mackenzie, Discussion of sea water sound–speed determinations. J. Acoust. Soc. Am. **70**(3), 801–806 (1981). https://doi.org/10.1121/1.386919
9. A.K. Mandal, S. Misra, M. Dash, Stochastic modeling of internal wave induced acoustic signal fluctuation and performance evaluation of shallow uwans, in *2013 IEEE International Conference on Communications Workshops (ICC)* (2013), pp. 1101–1105. https://doi.org/10.1109/ICCW.2013.6649401

10. S. Misra, A. Ghosh, The effects of variable sound speed on localization in underwater sensor networks, in *2011 Australasian Telecommunication Networks and Applications Conference (ATNAC)* (2011), pp. 1–4. https://doi.org/10.1109/ATNAC.2011.6096663
11. R. Pawlowicz, Key physical variables in the ocean: Temperature, salinity, and density. Nat. Educ. Knowl. **4**(4) (2013). https://www.nature.com/scitable/knowledge/library/key-physical-variables-in-the-ocean-temperature-102805293
12. U. Qureshi et al., RF path and absorption loss estimation for underwater wireless sensor networks in different water environments. Sensors **16**(6), 103–120 (2016). https://doi.org/10.3390/s16060890
13. A.H. Sayed, A. Tarighat, N. Khajehnouri, Network-based wireless location: challenges faced in developing techniques for accurate wireless location information. IEEE Signal Process. Mag. **22**(4), 24–40 (2005). https://doi.org/10.1109/MSP.2005.1458275
14. B. Szlachetko, M. Lower, Smart underwater positioning system and simultaneous communication, in *Computational Collective Intelligence* (Springer International Publishing, Cham, 2016), pp. 158–167
15. H. Tan, R. Diamant, W. Seah, M. Waldmeyer, A survey of techniques and challenges in underwater localization. Ocean Eng. **38**(14), 1663–1676 (2011). https://doi.org/10.1016/j.oceaneng.2011.07.017
16. The Newton–Laplace equation and speed of sound (2015). https://www.thermaxxjackets.com/newton-laplace-equation-sound-velocity/. Accessed 1 Jul 2020
17. A.L. Virovlyansky, A.Y. Kazarova, L.Y. Lyubavin, Ray-based description of normal modes in a deep ocean acoustic waveguide. J. Acoust. Soc. Am. **125**(3), 1362–1373 (2009). https://doi.org/10.1121/1.3075765

Chapter 5
ML: Modeling for Underwater Communication in IoUT Systems

5.1 Classification for Transmission Methods

In this section, we will consider predicting the suitable transmission method as a classification problem to select the desired class which represents the suitable transmission method. We will show an example of using the basic classification method which is the decision tree classifier; however, it will be applied on linear multihop transmission for IoUT system as in Fig. 5.1 [1].

Classification is one of the ML tasks; the objective is to predict the target class for each case of the dataset. Several classification approaches have been introduced to solve communication problems. Recently, classification techniques such as decision tree, rule-based method, memory-based learning, Bayesian networks, neural networks (NNs), and support vector machines have been used for numerous communication problems.

Decision tree (DT) is one of the most widely applied, fast, and efficient classification techniques. DT models require two phases, namely, the training and prediction phases. During the training phase, a sample dataset is used to form a tree model, while the model's output is compared with a target value in the prediction phase. The tree classification model is a decision support tool constructed from decision nodes, branches, and leaf nodes. Each decision node represents a test on an attribute, each branch represents the outcome of the test, and each leaf node represents a class label which represents the decision taken after computing all attributes. DT has considerable features and benefits as a classifier such as its simplicity in visualization and interpreting. In contrast to several machine learning approaches, DT requires minimal data preprocessing. The prediction time of DT is cost-effective and depends on data attributes. DT is able to deal with either categorical or numerical data.

Various decision trees splitting techniques have been introduced such as ID3 (Iterative Dichotomiser 3), C4.5, and CART. Building the DT is a recursive process based on the training dataset. The process first selects the most convenient attribute

A. A. Aziz El-Banna, K. Wu, *Machine Learning Modeling for IoUT Networks*,
SpringerBriefs in Computer Science, https://doi.org/10.1007/978-3-030-68567-6_5

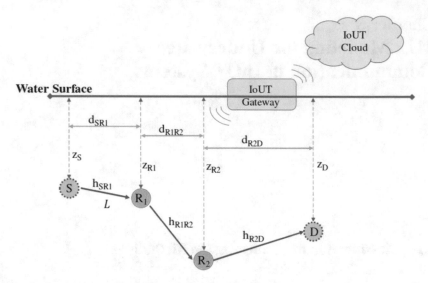

Fig. 5.1 Multihop cooperative transmission for an IoUT Network

and the best suitable value of that attribute. Both the selected attribute and its selected value should be the most effective pairs to split the tree for faster decisions. The entropy impurity is one of the effective measurement approaches to measure the splitting performance and is given as

$$i(N) = -\sum P(W_j) \log P(W_j) \tag{5.1}$$

where $P(W_j)$ is the fraction of the training dataset at a specific node that are in class W_j. The objective of the splitting is to decrease the impurity by δ, which is given as

$$\delta(i) = i(N) - P(x)i(N_x) - (1 - P(x))i(N_y) \tag{5.2}$$

where N_x and N_y are left and right nodes and $i(N_x)$ and $i(N_y)$ are their impurities. First, the impurity of each attribute is calculated and the highest one is then selected as the root node of the tree. Then, a splitting of the dataset into two sets left and right sub-trees is executed. Finally, a recursively start splitting of each sub-tree is executed.

The C4.5 [6] implementation of the decision trees is employed to train the model in [1]. The classification model is implemented by the attribute splitting, while the gain ratio impurity is used to select the splitting attributes. C4.5 is able to build a classification DT, also known as a statistical classifier [6], which is capable of handling continuous and discrete attributes and deal with noise and missing values.

The main objective of the aforementioned model is to take advantage of the ML techniques to formulate a new methodology that utilizes the data provided by simulations to predict the proper forwarding scheme based on the above algorithm.

Table 5.1 Classification confusion matrix for protocol selection

Predicted class	# AF selections	# DF selections
AF	933,587	336
DF	330	65,747

Table 5.2 Classification confusion matrix for fine-tuning the power level

Predicted class	Decrease P_{sound}	Increase P_{sound}
Decrease P_{sound}	499,830	0
Increase P_{sound}	0	500,170

The objective is achieved by employing the DT as a classifier with five feature vectors that represent the data corresponding to the values of T, pH, z, d, and f, which are used to build a model able to predict two restrictions, i.e., the forwarding protocol AF or DF and the appropriate amount of the signal intensity. DT uses hierarchical and sequential partitioning of data to find the most suitable classification model.

The size of the generated dataset is 1,000,000 samples with 90% of the data used as the training set and 10% as a testing set. Two DT models were built: the first model was to predict the protocol of choice and the second one to indicate whether the sound power should be increased or decreased according to the dynamic variation of the aforementioned environmental and system parameters. The employed models could classify data with remarkable accuracy, where the confusion matrix for the protocol selection DT is shown in Table 5.1 and its accuracy for training data is 99.85% and for testing 99.94% while the DT confusion matrix for fine-tuning the power level is shown in Table 5.2 with full mark model accuracy for both training and testing data.

5.2 Dynamic Modeling Using Neural Networks for Position Prediction

In this section, we will consider predicting the location of an underwater node as an application of ML to solve regression problems in underwater communication using dynamic modeling by neural networks. We start with an introduction to the different techniques that could be used to model nonlinear dynamic systems with NNs and then investigate the main differences among them. We then evaluate the dynamic NNs on the position prediction problem.

Artificial neural networks (ANNs) are divided into various branches; however, in this part, we examine the potential NNs for modeling the proposed approach. The widely known NN type is the feed-forward NN (FFNN) that has no loops between its units connections. The other common type that employs feedback connections within the network units is called recurrent NN (RNN). These feedback connections enable the RNN families to model sequential patterns. Other types (depicted in Fig. 5.2) of ANNs include radial basis function NNs, convolutional

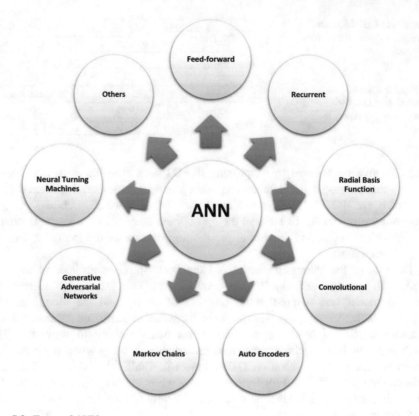

Fig. 5.2 Types of ANN

NNs, autoencoders, Markov chains, generative adversarial networks, neural turning machines, and others.

Modeling and solving various dynamic nonlinear problems could be achieved by using dynamic NNs (DyNets) that include tapped delay lines (TDLs) before different layers. However, the proper architecture of DyNets is determined based on the problem type. It is worth noting that DyNets and static networks, i.e., FFNN that has neither feedback elements nor TDLs, are another categorization for NNs. Moreover, we can classify dynamic nonlinear problems, or models, into three types as shown in Fig. 5.3.

The first type is the nonlinear autoregressive exogenous (NARX) modeling where the prediction of the required outputs is obtained by training the network using past samples from the input and output data [2]. The NARX model can be represented algebraically as

$$y_t = F(y_{t-1}, y_{t-2}, \ldots, y_{t-k_1}, u_{t-1}, u_{t-2}, \ldots, u_{t-k_2}) + \epsilon_t \qquad (5.3)$$

Fig. 5.3 Classification of
dynamic nonlinear problems

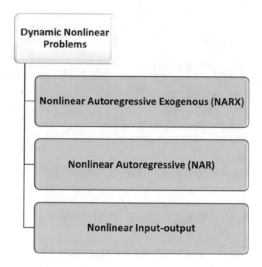

where y is the output or the variable of interest and u is the input or the externally
determined variable, ϵ_t is the error or noise term, and k_1 and k_2 are the TDL lengths
of the output and input data, respectively.

The second model is the nonlinear autoregressive (NAR), in which only past
samples from the output data are used to predict the output series. The NAR model
can be stated as

$$y_t = F(y_{t-1}, y_{t-2}, \ldots, y_{t-k_1}) + \epsilon_t \tag{5.4}$$

The final model is the nonlinear input–output, such as time delay NN (TDNN),
distributed delay NN (DDNN), and cascade-forward NN (CFNN), in which we
could predict the output series using only past samples from the input data
$(u_{t-1}, u_{t-2}, \ldots, u_{t-k_2})$. It should be mentioned that TDNN, DDNN, and CFNN
are considered non-recurrent DyNets, while NARX, NAR, and RNN are considered
recurrent DyNets.

This problem could be considered as a member of the third class, where the
KPVs (the inputs of the NN) are varying over the spatio-temporal domain; hence
past values of the KPV could help in predicting the location of the ROV (output of
the NN). As such, the predecessor output values are treated as of less importance;
therefore, their effect is ignored in that model.

Finally, Fig. 5.4 shows the architecture differences between various types of NN,
i.e., FFNN, RNN, TDNN, DDNN, and CFNN.

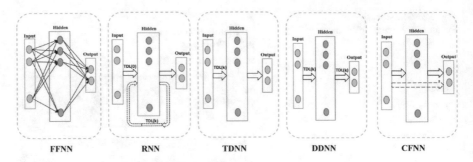

Fig. 5.4 Neural networks for dynamical modeling [3]

5.3 Performance Evaluation

5.3.1 Dataset and Preprocessing

The dataset used for training, testing, and validating the NNs is collected from two pools as follows [3].

- Firstly, the feature values of the environmental sensed amounts, i.e., T, S, z, ρ, etc., at the ROV place are drawn from the dataset in [5] which is collected from more than 20,000 separate archived datasets and employed numerous sets of measurements of the seawater physical variables vs. depth, e.g., temperature, salinity, and wind speed profiles.
- Secondly, the first part was employed as inputs to the proposed underwater positioning determination approach in the Algorithms shown in Figs. 4.2 and 4.4 in Chap. 4 to obtain the labels (i.e., estimated locations) of the ROV for the TDoA- and RSS-based positioning techniques, respectively.
- Finally, these values are combined to formulate around 10K sample dataset that contains the complete features and labels for the NN as follows:
  - *twenty-one features* consist of the KPV values T, S, z, ρ, θ, and p at the location of anchor 1, anchor 2, and the ROV plus the entire distances between them, i.e., d, r_1, and r_2.
  - *two labels* that represent the required location of the ROV, i.e., $[x_r; y_r]$.

Moreover, we applied feature normalization using the min–max normalization technique to rescale the data range to [0,1] for proper learning of the different employed NNs. In addition, the "preparets" built-in Matlab function is employed to prepare the input and target data for network training by shifting the input and target samples by certain steps k_1 and k_2 to fill the initial input and layer delay states.

Finally, hyperparameter tuning is performed using grid search over the number of neurons (hidden layer size) to get the best value that doesn't degrade the performance while avoiding extra computational complexity (to enhance the node lifetime since most underwater nodes are battery-based devices).

5.3.2   Training and Evaluating the DyNets

Making use of the aforementioned dataset, the Matlab environment is used with the settings in training the NNs mentioned in Fig. 5.4 as shown in Table 5.3.

To evaluate the training performance of the different Dynets, a study on the training time is performed, and the results are shown in Fig. 5.5 using three sets of neurons in the hidden layer ($N = 2$, $N = 4$, and $N = 6$). As shown from Fig. 5.5, the RNN and CFNN required large execution time to operate; therefore, they were avoided from being proper DyNets for the underwater nodes to save the battery lifetime of these devices. Normalized to the FFNN training time, the RNN and CFNN required around 110 and 15 multiple execution times than the FFNN. In addition, the other networks are evaluated where the DDNN requires around only

Table 5.3 Matlab simulation settings for NN

Data division	70% training
	15% testing
	15% validation
Training algorithm	Levenberg–Marquardt back propagation (LMBP)
Performance loss function	Mean squared normalized error (MSE)
Simulation and gradient calculations	Minimum excluded (MEX)
Epoch/mu	1000/0.001

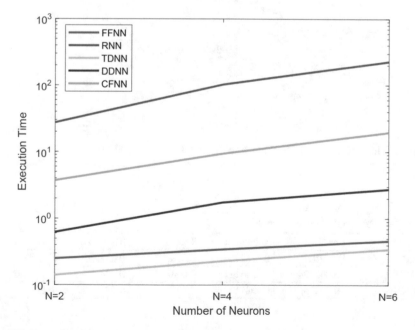

Fig. 5.5 Execution time evaluation for different DyNets

twice the execution time of the FFNN, while the TDNN requires only half the execution time of the FFNN.

The three networks, i.e., FFNN, TDNN, and DDNN, are then evaluated in terms of the mean squared normalized error (MSE) as a measure for their performance in the training, validation, and testing phases.

Figure 5.6 shows the performance of the best networks for the FFNN, TDNN, and DDNN in terms of the best validation (Bval) MSE values and their distributions along the 100 initialization. The three networks show a comparable performance with Bval mean MSE of $\approx 2 \times 10^{-3}$; moreover, they show almost the same level of stability. Other performance details calculated over all the 100 simulation samples such as the training, testing, validation performances, and the execution time are shown in Table 5.4.

From the above results, it is shown that the FFNN, TDNN, DDNN show a comparable performance in the three phases of training, testing, and validation and can model the underwater positioning systems with small errors (around $-47\,$dB). However, the FFNN outperforms the two others from the accuracy perspective (i.e., has lower MSE values), while the TDNN requires the lowest execution time and hence is the best DyNets from the energy consumption perspective.

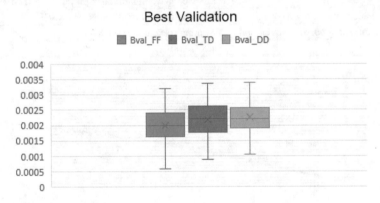

Fig. 5.6 The best validation (Bval) of the FFNN, TDNN, and DDNN

Table 5.4 DyNets performance metrics

Performance	Bval	Validation	Test	Train	Time
FFNN	0.001996	0.018276	0.018253	0.018252	0.250771
TDNN	0.002194	0.034411	0.034472	0.034135	0.141032
DDNN	0.002256	0.038115	0.038062	0.037999	0.623007

References

1. A.A. Aziz El-Banna, A.B. Zaky, B.M. ElHalawany, J. Zhexue Huang, K. Wu, Machine learning based dynamic cooperative transmission framework for IoUT networks, in *2019 16th Annual IEEE International Conference on Sensing, Communication, and Networking (SECON)* (2019), pp. 1–9
2. S.A. Billingsg, *Nonlinear System Identification: NARMAX Methods in the Time, Frequency, and Spatio-Temporal Domains* (Wiley, Chichester, 2013)
3. A.A.A. El-Banna, K. Wu, B.M. ElHalawany, Application of neural networks for dynamic modeling of an environmental-aware underwater acoustic positioning systems using seawater physical properties. IEEE Geosci. Remote Sens. Lett. (2020). https://ieeexplore.ieee.org/document/9286842
4. Q. Mao, F. Hu, Q. Hao, Deep learning for intelligent wireless networks: A comprehensive survey. IEEE Commun. Surv. Tutorials **20**(4), 2595–2621 (2018)
5. NCEI standard product: World ocean database (WOD). https://catalog.data.gov/dataset/ncei-standard-product-world-ocean-database-wod
6. R. Quinlan, *4.5: Programs for Machine Learning* (Morgan Kaufmann Publishers, San Francisco, 1993)
7. T. Wang, C.K. Wen, H. Wang, F. Gao, T. Jiang, S. Jin, Deep learning for wireless physical layer: Opportunities and challenges. China Commun. **14**(11), 92–111 (2017)

Chapter 6
Open Challenges for IoUT Networks

6.1 Communication Challenges

There are various challenges that face the design of underwater acoustic networks [1]. The most serious challenge that limits the operation of various underwater techniques is the underwater media that are severely impaired due to its harsh nature that increases multi-path and fading effects. In addition, underwater channels suffer also from having a very high propagation delay than terrestrial channels, and it's very difficult to have a standard model for them. Moreover, as a result of these channel characteristics, high bit error rates and temporary losses of connectivity usually occur. That is besides the limited bandwidth of the acoustic communication that usually suits the underwater.

Another group of challenges that are related to the underwater sensor nodes is battery power that usually couldn't be recharged and being highly prone to failures because of fouling and corrosion.

In the following, we briefly discuss the challenges of media characteristics, channel modeling, and complexity of the hardware and other combined technologies.

6.1.1 Media Characteristics

As mentioned above, the major challenge that faces underwater networking is the characteristics of the transmission media. Therefore, selecting a suitable carrier is a very hard task. The proper candidates for underwater transmission are limited to the widely used acoustic signals, while RF and optical transmissions are impractical in many scenarios.

Radio signals with high frequencies are exposed to absorption and high attenuation levels when traveling underwater. However, at low frequencies, they require high transmission power to propagate for long distance. On the other hand, optical

© The Author(s), under exclusive license to Springer Nature Switzerland AG 2021
A. A. Aziz El-Banna, K. Wu, *Machine Learning Modeling for IoUT Networks*,
SpringerBriefs in Computer Science, https://doi.org/10.1007/978-3-030-68567-6_6

signals usually could propagate for a very short distance in underwater environments as they are exposed to scattering and high absorption which limits their application for underwater transmission as well. However, it should be noted that underwater wireless optical communication technology has developed rapidly in recent years to provide feasible transmission frameworks.

Acoustic signals with lower attenuation and absorption can propagate much longer distances (than both the radio and optical signals) in underwater environments [4]. However, acoustic signals are the proper solution, but we still need to consider the dynamics of the underwater environment and their effect on sound transmissions.

6.1.2   Underwater Channel Modeling

One of the solutions to overcome the compensation between the extended coverage range and high data rates is to employ flexible transmissions that could be achieved by the assistance of non-acoustic underwater transmission systems. This is known as a multimodal transmission which aims to incorporate different physical communication technologies [2]. This technology could compensate for the drawbacks of one technology through the advantages of another. As such, a hybrid transmission that used acoustic and optical signals could achieve long-range, low data rate communications as well as fast data transmission at short ranges [2].

Although this technology faces up many challenges, the main challenge is the underwater channel modeling. Accurate channel estimations are required to determine the proper transmission frequency and power level of the underwater nodes. Studies in this area should include a joint combination of vertical and horizontal underwater optical links and the development of 3D magnetic induction underwater channel modeling that jointly considers power allocations and coding rates[5].

Moreover, systems and devices that employ multimodal technology will be more complex and expensive when employing different physical communication technologies on the same chip.

6.1.3   Complexity of the Hardware and Other Combined Technologies

The complexity in the system comes from many directions. One direction is the computational complexity of the algorithms needed to handle multimodal transmission and other adaptive parameters. The second one is the physical hardware components and circuits required to combine all the modules and different layers of communication and sensing. For example, in localization, as the number of anchors and nodes becomes large, the quantity of ToA or TDoA measurements highly increases, and as a result, the computational complexity will be very high. In addition, higher complexity will cause additional delays besides consuming more resources.

Contributions on low-complexity algorithms as well as smart designs of physical components are still needed, and more efforts from the researchers are required in this direction.

6.2 ML Challenges

Traditional ML algorithms face the following challenges[3]: insufficient quantity, non-representative, and poor quality of training data; irrelevant features; overfitting and underfitting the training data; and the black box nature of some of its techniques such as NNs. This is besides the requirements of talented developers and the high computational power demands to train the models very well.

In the following, we briefly shed the light on the challenges of availability of datasets, problem formulation and model selection, the feasibility of online learning, and the importance of developing reduced computational complexity models.

6.2.1 Datasets Availability

We can say that the "dataset is the core of the model." To design and train a convenient model for a certain problem, one needs thousands to millions of training examples depending on the problem size. In addition, to achieve good generalization, the training data must be representative enough to encompass new cases of unseen data. Furthermore, poor-quality measurements produce data full of errors and outliers, which makes the recognition of the underlying patterns very difficult.

Therefore, the fundamental challenge for applying machine learning is the lack of datasets that accurately include all the underwater environments as well as combining all needed physical variables and also being recorded all over the year to cover different operating scenarios. However, many datasets are available, but these datasets are either incomplete or cover only specific underwater regions, which complicates the process of developing generic models.

Some solutions that may give a hand in solving this problem are using the techniques of data augmentation, ensemble learning, or combining experimental and simulated datasets to produce more robust models that best describe the underwater problems. Also feature engineering could help in producing pretty sets of features for training. Feature engineering is usually following three steps that are feature selection, feature extraction, and then making new features by combining new data.

6.2.2   Problem Formulation and Model Selection

Although it is considered a common challenge for all the applications of machine learning, formulating the problem and accurately selecting the best model are very difficult tasks and typically take a long process and time. Tens of machine learning models exist, and most of them work fine with many problems, but still, other models can perform better from the perspective of the performance and the reduced complexity, and the latter is a very critical matter where underwater devices almost have no chances for recharging which makes the power saving a critical issue.

6.2.3   Feasibility of Online Learning

Once training a model, we sometimes find it forget what it learned while trying to train it on another task. In that case, we still need the model to learn during operation to adapt itself based on the new tasks and data, and this is the core idea of online learning. Online learning is the basic form of gradual learning which is an active area of research.

Online learning is extremely hard and could be a challenge specifically with the reduced power capability of underwater devices. Since for online learning, we typically have more data, but we have time constraints and limited power capabilities; these issues should be addressed in more detail to make use of these vast data to enhance the learning capabilities of the designed models and obtain rigid models valid for various operating scenarios.

6.2.4   Reduced Computational Complexity Models

Considering underwater devices as power-limited devices or battery-based motes and they are not usually able to recharge, all developed ML models, as well as communication algorithms, should be very simple, and their computational complexity should be reduced. This opens the door for researchers to design and develop new ML techniques that have low computational complexity that suits underwater communication and IoUT.

References

1. I.F. Akyildiz, D. Pompili, T. Melodia, Underwater acoustic sensor networks: research challenges. Ad Hoc Netw. **3**(3), 257–279 (2005)
2. F. Campagnaro, R. Francescon, P. Casari, R. Diamant, M. Zorzi, Multimodal underwater networks: Recent advances and a look ahead, in *Proceedings of the International Conference on Underwater Networks & Systems* (2017), pp. 1–8

3. A. Géron, *Hands-on machine learning with Scikit-Learn, Keras, and TensorFlow: Concepts, tools, and techniques to build intelligent systems* (O'Reilly Media, Newton, 2019)
4. E.C. Liou, C.C. Kao, C.H. Chang, Y.S. Lin, C.J. Huang, Internet of underwater things: Challenges and routing protocols, in *2018 IEEE International Conference on Applied System Invention (ICASI)* (IEEE, Piscataway, 2018), pp. 1171–1174
5. R. Su, D. Zhang, C. Li, Z. Gong, R. Venkatesan, F. Jiang, Localization and data collection in auv-aided underwater sensor networks: Challenges and opportunities. IEEE Netw. **33**(6), 86–93 (2019)

Index

© The Author(s), under exclusive license to Springer Nature Switzerland AG 2021
A. A. Aziz El-Banna, K. Wu, *Machine Learning Modeling for IoUT Networks*,
SpringerBriefs in Computer Science, https://doi.org/10.1007/978-3-030-68567-6

Printed in the United States
by Baker & Taylor Publisher Services